T0332882

CAMBRIDGE TRACTS IN MATHEMATICS

General Editors

B. BOLLOBÁS, W. FULTON, A. KATOK, F. KIRWAN, P. SARNAK, B. SIMON, B. TOTARO

---

198 Topics in Critical Point Theory

CAMBRIDGE TRACTS IN MATHEMATICS

GENERAL EDITORS

B. BOLLOBÁS, W. FULTON, A. KATOK, F. KIRWAN, P. SARNAK, B. SIMON, B. TOTARO

A complete list of books in the series can be found at www.cambridge.org/mathematics.
Recent titles include the following:

Topics in Critical Point Theory

KANISHKA PERERA

Florida Institute of Technology

MARTIN SCHECHTER

University of California, Irvine

CAMBRIDGE
UNIVERSITY PRESS

Shaftesbury Road, Cambridge CB2 8EA, United Kingdom

One Liberty Plaza, 20th Floor, New York, NY 10006, USA

477 Williamstown Road, Port Melbourne, VIC 3207, Australia

314–321, 3rd Floor, Plot 3, Splendor Forum, Jasola District Centre, New Delhi – 110025, India

103 Penang Road, #05–06/07, Visioncrest Commercial, Singapore 238467

Cambridge University Press is part of Cambridge University Press & Assessment, a department of the University of Cambridge.

We share the University's mission to contribute to society through the pursuit of education, learning and research at the highest international levels of excellence.

www.cambridge.org
Information on this title: www.cambridge.org/9781107029668

© Kanishka Perera and Martin Schechter 2013

First published 2013

A catalogue record for this publication is available from the British Library

Library of Congress Cataloging-in-Publication data
Perera, Kanishka, 1969–
Topics in critical point theory / Kanishka Perera, Florida Institute of Technology; Martin Schechter, University of California, Irvine.
pages   cm. – (Cambridge tracts in mathematics ; 198)
Includes bibliographical references and index.
ISBN 978-1-107-02966-8
1. Fixed point theory.   I. Schechter, Martin.   II. Title.
QA329.9.P47   2013
514'.74 – dc23   2012025065

ISBN   978-1-107-02966-8   Hardback

To my wife Champa.
           K.P.

To my wife, Deborah, our children,
our grandchildren (twenty five so far),
our great grandchildren (fifteen so far),
and our extended family.
May they all enjoy many happy years.
           M.S.

Contents

Preface

Critical point theory has become a very powerful tool for solving many problems. The theory has enjoyed significant development over the past several years. The impetus for this development is the fact that many new problems could not be solved by the older theory.

There have been several excellent books written on critical point theory from various points of view; see, e.g., Berger [19], Zeidler [161], Rabinowitz [129], Mawhin and Willem [91], Chang [29, 30], Ghoussoub [56], Ambrosetti and Prodi [8], Willem [158], Chabrowski [26], Dacorogna [36], and Struwe [153] (see also Schechter [143, 144, 147], Zou and Schechter [163], and Perera *et al.* [113]). In this book we present more recent developments in the subject that do not seem to be covered elsewhere, including some results of the authors dealing with nonstandard linking geometries and sandwich pairs.

Chapter 1 is a brief review of Morse theory in Banach spaces. We prove the first and second deformation lemmas under the Cerami compactness condition. As the variational functionals associated with applications given later in the book will only be C^1, we discuss critical groups of C^1-functionals. We include discussions on minimizers, nontrivial critical points, mountain pass points, and the three critical points theorem. We also give a generalized notion of local linking that yields a nontrivial critical group, which will be applied to problems with jumping nonlinearities in Chapter 5. We close the chapter with a recent result of Perera [110] on nontrivial critical groups in p-Laplacian problems.

Chapter 2 is on linking. We say that subsets A, B of a Banach space E link if every C^1-functional G on E satisfying

$$-\infty < a := \sup_A G \leqslant \inf_B G =: b < +\infty,$$

and a suitable compactness condition, has a critical point u with $G(u) \geqslant b$. There are three main notions of linking; homological, homotopical, and a more recent one introduced by Schechter and Tintarev [148]. We discuss all three and show that

$$\text{homological} \atop \text{linking} \quad \Longrightarrow \quad \text{homotopical} \atop \text{linking} \quad \Longrightarrow \quad \text{Schechter–Tintarev} \atop \text{linking.}$$

We also discuss some results of Schechter and Tintarev [148] on pairs of critical points produced by linking subsets and some results on their critical groups due to Perera [104]. We close with some recent results of the authors [118] that give critical points with nontrivial critical groups under nonstandard geometrical assumptions that do not involve a finite-dimensional closed loop.

Chapter 3 contains applications of the Morse theoretic and linking methods of the first two chapters to semilinear elliptic boundary value problems. We discuss the local nature of critical groups and critical groups at zero. We consider asymptotically linear problems and problems with concave nonlinearities, and obtain multiple nontrivial solutions using our nonstandard linking theorems.

Chapter 4 considers the Fučík spectrum in an abstract operator setting, that includes many concrete problems arising in applications as special cases. We construct the minimal and maximal curves of the spectrum locally near the points where it intersects the main diagonal of the plane. We give a sufficient condition for the region between them to be nonempty, and show that it is free of the spectrum in the case of a simple eigenvalue. Finally we compute the critical groups in various regions separated by these curves. We compute them precisely in Type I regions, and prove a shifting theorem that gives a finite-dimensional reduction to the null manifold for Type II regions.

Chapter 5 is a continuation of the previous chapter that considers problems with jumping nonlinearities in the same abstract framework. We discuss compactness and critical groups at infinity and zero. We compute critical groups in both resonant and nonresonant problems. This allows us to establish solvability in Type I regions, and obtain nontrivial solutions for nonlinearities crossing a curve of the Fučík spectrum constructed in Chapter 4.

Chapter 6 is on sandwich pairs. We say that a pair of subsets A, B of a Banach space E is a sandwich pair if every C^1-functional G on E satisfying

$$-\infty < b := \inf_B G \leqslant \sup_A G =: a < +\infty,$$

and a suitable compactness condition, has a critical point u with $b \leqslant G(u) \leqslant a$. We construct a very general class of sandwich pairs with wide applicability using a family of flows on E and the Fadell–Rabinowitz cohomological index. We use a special case of this where the sandwich pair is a certain pair of cones to solve p-Laplacian problems. We also solve anisotropic p-Laplacian systems using another special case with a curved sandwich pair made up of certain orbits of an associated group action on a product of Sobolev spaces.

1

Morse theory

1.1 Introduction

The purpose of this chapter is to introduce the reader to Morse theoretic methods used in variational problems. General references are Milnor [93], Mawhin and Willem [91], Chang [29], and Benci [17]; see also Perera *et al.* [113]. We begin by briefly collecting some basic results of Morse theory. These include the Morse inequalities, Morse lemma and its generalization splitting lemma, the shifting theorem of Gromoll and Meyer, and the handle body theorem. Results that are needed later in the text will be proved in subsequent sections.

Let G be a real-valued function defined on a real Banach space E. We say that G is Fréchet differentiable at $u \in E$ if there is an element $G'(u)$ of the dual E', called the Fréchet derivative of G at u, such that

$$G(u + v) = G(u) + (G'(u), v) + o(\|v\|) \text{ as } v \to 0 \text{ in } E,$$

where $(\cdot, \cdot)$ is the duality pairing. The functional G is continuously Fréchet differentiable on E, or belongs to the class $C^1(E, \mathbb{R})$, if G' is defined everywhere and the map $E \to E'$, $u \mapsto G'(u)$ is continuous. We assume that $G \in C^1(E, \mathbb{R})$ for the rest of the chapter. Replacing G with $G - G(0)$ if necessary, we may also assume that $G(0) = 0$. The functional G is called even if

$$G(-u) = G(u) \quad \forall u \in E.$$

Then G' is odd, i.e.,

$$G'(-u) = -G'(u) \quad \forall u \in E.$$

We say that u is a critical point of G if $G'(u) = 0$. A value c of G is a critical value if there is a critical point u with $G(u) = c$, otherwise it is a regular value.

We use the standard notations

$$G_a = \{u \in E : G(u) \geqslant a\}, \qquad G^b = \{u \in E : G(u) \leqslant b\},$$
$$G_a^b = G_a \cap G^b,$$
$$K = \{u \in E : G'(u) = 0\}, \qquad \hat{E} = E \backslash K,$$
$$K_a^b = K \cap G_a^b, \qquad K^c = K_c^c$$

for the various superlevel, sublevel, critical, and regular sets of G.

It is usually necessary to assume some sort of a "compactness condition" when seeking critical points of a functional. The following condition was originally introduced by Palais and Smale [101]: G satisfies the Palais–Smale compactness condition at the level c, or $(\text{PS})_c$ for short, if every sequence $(u_j) \subset E$ such that

$$G(u_j) \to c, \quad G'(u_j) \to 0,$$

called a $(\text{PS})_c$ sequence, has a convergent subsequence; G satisfies (PS) if it satisfies $(\text{PS})_c$ for every $c \in \mathbb{R}$, or equivalently, if every sequence such that $G(u_j)$ is bounded and $G'(u_j) \to 0$, called a (PS) sequence, has a convergent subsequence. The following weaker version was introduced by Cerami [25]: G satisfies the Cerami condition at the level c, or $(\text{C})_c$ for short, if every sequence such that

$$G(u_j) \to c, \quad \left(1 + \|u_j\|\right) G'(u_j) \to 0,$$

called a $(\text{C})_c$ sequence, has a convergent subsequence; G satisfies (C) if it satisfies $(\text{C})_c$ for every c, or equivalently, if every sequence such that $G(u_j)$ is bounded and $\left(1 + \|u_j\|\right) G'(u_j) \to 0$, called a (C) sequence, has a convergent subsequence. This condition is weaker since a $(\text{C})_c$ (resp. (C)) sequence is clearly a $(\text{PS})_c$ (resp. (PS)) sequence also. The limit of a $(\text{PS})_c$ (resp. (PS)) sequence is in K^c (resp. K) since G and G' are continuous. Since any sequence in K^c is a $(\text{C})_c$ sequence, it follows that K^c is a compact set when $(\text{C})_c$ holds.

Some of the essential tools for locating critical points are the deformation lemmas, which allow to lower sublevel sets of a functional, away from its critical set. The main ingredient in their proofs is a suitable negative pseudo-gradient flow, a notion due to Palais [103]: a pseudo-gradient vector field for G on $\hat{E}$ is a locally Lipschitz continuous mapping $V : \hat{E} \to E$ satisfying

$$\|V(u)\| \leqslant \|G'(u)\|, \quad 2\left(G'(u), V(u)\right) \geqslant \left(\|G'(u)\|\right)^2 \quad \forall u \in \hat{E}.$$

Such a vector field exists, and may be chosen to be odd when G is even.

The first deformation lemma provides a local deformation near a (possibly critical) level set of a functional.

Lemma 1.1.1 (first deformation lemma) *If $c \in \mathbb{R}$, C is a bounded set containing K^c, $\delta, k > 0$, and G satisfies $(C)_c$, then there are an $\varepsilon_0 > 0$ and, for each $\varepsilon \in (0, \varepsilon_0)$, a map $\eta \in C(E \times [0, 1], E)$ satisfying*

(i) $\eta(\cdot, 0) = id_E$,
(ii) $\eta(\cdot, t)$ *is a homeomorphism of E for all $t \in [0, 1]$,*
(iii) $\eta(\cdot, t)$ *is the identity outside $A = G_{c-2\varepsilon}^{c+2\varepsilon} \backslash N_{\delta/3}(C)$ for all $t \in [0, 1]$,*
(iv) $\|\eta(u, t) - u\| \leqslant (1 + \|u\|)\, \delta/k \quad \forall (u, t) \in E \times [0, 1]$,
(v) $G(\eta(u, \cdot))$ *is nonincreasing for all $u \in E$,*
(vi) $\eta(G^{c+\varepsilon} \backslash N_\delta(C), 1) \subset G^{c-\varepsilon}$.

When G is even and C is symmetric, η may be chosen so that $\eta(\cdot, t)$ is odd for all $t \in [0, 1]$.

The first deformation lemma under the $(PS)_c$ condition is due to Palais [102]; see also Rabinowitz [126]. The proof under the $(C)_c$ condition was given by Cerami [25] and Bartolo *et al.* [13]. The particular version given here will be proved in Section 1.3.

The second deformation lemma implies that the homotopy type of sublevel sets can change only when crossing a critical level.

Lemma 1.1.2 (second deformation lemma) *If $-\infty < a < b \leqslant +\infty$ and G has only a finite number of critical points at the level a, has no critical values in (a, b), and satisfies $(C)_c$ for all $c \in [a, b] \cap \mathbb{R}$, then G^a is a deformation retract of $G^b \backslash K^b$, i.e. there is a map $\eta \in C((G^b \backslash K^b) \times [0, 1], G^b \backslash K^b)$, called a deformation retraction of $G^b \backslash K^b$ onto G^a, satisfying*

(i) $\eta(\cdot, 0) = id_{G^b \backslash K^b}$,
(ii) $\eta(\cdot, t)|_{G^a} = id_{G^a} \quad \forall t \in [0, 1]$,
(iii) $\eta(G^b \backslash K^b, 1) = G^a$.

The second deformation lemma under the $(PS)_c$ condition is due to Rothe [135], Chang [28], and Wang [157]. The proof under the $(C)_c$ condition can be found in Bartsch and Li [14], Perera and Schechter [119], and in Section 1.3.

In Morse theory the local behavior of G near an isolated critical point u is described by the sequence of critical groups

$$C_q(G, u) = H_q(G^c \cap U, G^c \cap U \backslash \{u\}), \quad q \geqslant 0$$

where $c = G(u)$ is the corresponding critical value, U is a neighborhood of u, and H_* denotes singular homology. They are independent of the choice of U by the excision property.

For example, if u is a local minimizer, $C_q(G, u) = \delta_{q0}\mathcal{G}$ where δ is the Kronecker delta and $\mathcal{G}$ is the coefficient group. A critical point u with $C_1(G, u) \neq 0$ is called a mountain pass point.

Let $-\infty < a < b \leqslant +\infty$ be regular values and assume that G has only isolated critical values $c_1 < c_2 < \cdots$ in (a, b), with a finite number of critical points at each level, and satisfies (PS)$_c$ for all $c \in [a, b] \cap \mathbb{R}$. Then the Morse type numbers of G with respect to the interval (a, b) are defined by

$$M_q(a, b) = \sum_i \operatorname{rank} H_q(G^{a_{i+1}}, G^{a_i}), \quad q \geqslant 0$$

where $a = a_1 < c_1 < a_2 < c_2 < \cdots$. They are independent of the a_i by the second deformation lemma, and are related to the critical groups by

$$M_q(a, b) = \sum_{u \in K_a^b} \operatorname{rank} C_q(G, u).$$

Writing $\beta_j(a, b) = \operatorname{rank} H_j(G^b, G^a)$, we have the following.

Theorem 1.1.3 (Morse inequalities)   *If there is only a finite number of critical points in G_a^b, then*

$$\sum_{j=0}^{q} (-1)^{q-j} M_j \geqslant \sum_{j=0}^{q} (-1)^{q-j} \beta_j, \quad q \geqslant 0,$$

and

$$\sum_{j=0}^{\infty} (-1)^j M_j = \sum_{j=0}^{\infty} (-1)^j \beta_j$$

when the series converge.

Critical groups are invariant under homotopies that preserve the isolatedness of the critical point; see Rothe [134], Chang and Ghoussoub [27], and Corvellec and Hantoute [32].

Theorem 1.1.4   *If G_t, $t \in [0, 1]$ is a family of C^1-functionals on E satisfying (PS), u is a critical point of each G_t, and there is a closed neighborhood U such that*

(i) *U contains no other critical points of G_t,*
(ii) *the map $[0, 1] \to C^1(U, \mathbb{R})$, $t \mapsto G_t$ is continuous,*

then $C_*(G_t, u)$ are independent of t.

When the critical values are bounded from below and G satisfies (C), the global behavior of G can be described by the critical groups at infinity introduced by Bartsch and Li [14]

$$C_q(G, \infty) = H_q(E, G^a), \quad q \geqslant 0$$

where a is less than all critical values. They are independent of a by the second deformation lemma and the homotopy invariance of the homology groups.

For example, if G is bounded from below, $C_q(G, \infty) = \delta_{q0} \mathcal{G}$. If G is unbounded from below, $C_q(G, \infty) = \tilde{H}_{q-1}(G^a)$ where $\tilde{H}$ denotes the reduced groups.

Proposition 1.1.5 *If $C_q(G, \infty) \neq 0$ and G has only a finite number of critical points and satisfies* (C), *then G has a critical point u with $C_q(G, u) \neq 0$.*

The second deformation lemma implies that $C_q(G, \infty) = C_q(G, 0)$ if $u = 0$ is the only critical point of G, so G has a nontrivial critical point if $C_q(G, 0) \neq C_q(G, \infty)$ for some q.

Now suppose that E is a Hilbert space $(H, (\cdot, \cdot))$ and $G \in C^2(H, \mathbb{R})$. Then the Hessian $A = G''(u)$ is a self-adjoint operator on H for each u. When u is a critical point the dimension of the negative space of A is called the Morse index of u and is denoted by $m(u)$, and $m^*(u) = m(u) + \dim \ker A$ is called the large Morse index. We say that u is nondegenerate if A is invertible. The Morse lemma describes the local behavior of the functional near a nondegenerate critical point.

Lemma 1.1.6 (Morse lemma) *If u is a nondegenerate critical point of G, then there is a local diffeomorphism ξ from a neighborhood U of u into H with $\xi(u) = 0$ such that*

$$G(\xi^{-1}(v)) = G(u) + \frac{1}{2}(Av, v), \quad v \in \xi(U).$$

Morse lemma in $\mathbb{R}^n$ was proved by Morse [95]. Palais [102], Schwartz [149], and Nirenberg [98] extended it to Hilbert spaces when G is C^3. Proof in the C^2 case is due to Kuiper [64] and Cambini [23].

A direct consequence of the Morse lemma is the following theorem.

Theorem 1.1.7 *If u is a nondegenerate critical point of G, then*

$$C_q(G, u) = \delta_{qm(u)} \mathcal{G}.$$

The handle body theorem describes the change in topology as the level sets pass through a critical level on which there are only nondegenerate critical points.

Theorem 1.1.8 (handle body theorem) *If c is an isolated critical value of G for which there are only a finite number of nondegenerate critical points u_i, $i = 1, \ldots, k$, with Morse indices $m_i = m(u_i)$, and G satisfies* (PS), *then there are an $\varepsilon > 0$ and homeomorphisms φ_i from the unit disk D^{m_i} in $\mathbb{R}^{m_i}$ into H such that*

$$G^{c-\varepsilon} \cap \varphi_i(D^{m_i}) = G^{-1}(c - \varepsilon) \cap \varphi_i(D^{m_i}) = \varphi_i(\partial D^{m_i})$$

and $G^{c-\varepsilon} \cup \bigcup_{i=1}^{k} \varphi_i(D^{m_i})$ is a deformation retract of $G^{c+\varepsilon}$.

The references for Theorems 1.1.3, 1.1.7, and 1.1.8 are Morse [95], Pitcher [124], Milnor [93], Rothe [132, 133, 135], Palais [102], Palais and Smale [101], Smale [151], Marino and Prodi [89], Schwartz [149], Mawhin and Willem [91], and Chang [29].

The splitting lemma generalizes the Morse lemma to degenerate critical points. Assume that the origin is an isolated degenerate critical point of G and 0 is an isolated point of the spectrum of $A = G''(0)$. Let $N = \ker A$ and write $H = N \oplus N^{\perp}$, $u = v + w$.

Lemma 1.1.9 (splitting lemma) *There are a ball $B \subset H$ centered at the origin, a local homeomorphism ξ from B into H with $\xi(0) = 0$, and a map $\eta \in C^1(B \cap N, N^{\perp})$ such that*

$$G(\xi(u)) = \frac{1}{2}(Aw, w) + G(v + \eta(v)), \quad u \in B.$$

Splitting lemma when A is a compact perturbation of the identity was proved by Gromoll and Meyer [57] for $G \in C^3$ and by Hofer [60] in the C^2 case. Mawhin and Willem [90, 91] extended it to the case where A is a Fredholm operator of index zero. The general version given here is due to Chang [29].

A consequence of the splitting lemma is the following.

Theorem 1.1.10 (shifting theorem) *We have*

$$C_q(G, 0) = C_{q-m(0)}(G|_{\mathcal{N}}, 0) \quad \forall q$$

where $\mathcal{N} = \xi(B \cap N)$ is the degenerate submanifold of G at 0.

The shifting theorem is due to Gromoll and Meyer [57]; see also Mawhin and Willem [91] and Chang [29].

Since $\dim \mathcal{N} = m^*(0) - m(0)$, the shifting theorem gives us the following Morse index estimates when there is a nontrivial critical group.

Corollary 1.1.11 *If $C_q(G, 0) \neq 0$, then*

$$m(0) \leqslant q \leqslant m^*(0).$$

It also enables us to compute the critical groups of a mountain pass point of nullity at most one.

Corollary 1.1.12 *If u is a mountain pass point of G and* dim ker $G''(u) \leqslant 1$, *then*

$$C_q(G, u) = \delta_{q1} \mathcal{G}.$$

This result is due to Ambrosetti [4, 5] in the nondegenerate case and to Hofer [60] in the general case.

Shifting theorem also implies that all critical groups of a critical point with infinite Morse index are trivial, so the above theory is not suitable for studying strongly indefinite functionals. An infinite-dimensional Morse theory particularly well suited to deal with such functionals was developed by Szulkin [155]; see also Kryszewski and Szulkin [63].

The following important perturbation result is due to Marino and Prodi [88]; see also Solimini [152].

Theorem 1.1.13 *If some critical value of G has only a finite number of critical points u_i, $i = 1, \ldots, k$ and $G''(u_i)$ are Fredholm operators, then for any sufficiently small $\varepsilon > 0$ there is a C^2-functional G_ε on H such that*

(i) $\|G_\varepsilon - G\|_{C^2(H)} \leqslant \varepsilon$,

(ii) $G_\varepsilon = G$ in $H \setminus \bigcup_{i=1}^k B_\varepsilon(u_j)$,

(iii) G_ε has only nondegenerate critical points in $B_\varepsilon(u_j)$ and their Morse indices are in $[m(u_i), m^*(u_i)]$,

(iv) G satisfies (PS) $\implies$ G_ε satisfies (PS).

Here

$$B_r(u_0) = \{u \in H : \|u - u_0\| \leqslant r\}$$

is the closed ball of radius r centered at u_0. We will write B_r for $B_r(0)$ in the sequel.

Returning to the setting of a C^1-functional on a Banach space E, in many applications G has the trivial critical point $u = 0$ and we are interested in finding others. The notion of a local linking introduced by Li and Liu [70, 83] is useful for obtaining nontrivial critical points under various assumptions on the behavior of G at infinity; see also Brezis and Nirenberg [21] and Li and Willem [72]. Assume that the origin is a critical point of G with $G(0) = 0$. We say that G has a local linking near the origin if there is a direct sum decomposition $E = N \oplus M$, $u = v + w$ with N finite dimensional such that

$$
\begin{cases}
G(v) \leqslant 0, & v \in N, \ \|v\| \leqslant r \\
G(w) > 0, & w \in M, \ 0 < \|w\| \leqslant r
\end{cases}
$$

for sufficiently small $r > 0$. Liu [82] showed that this yields a nontrivial critical group at the origin.

Proposition 1.1.14 *If G has a local linking near the origin with $\dim N = d$ and the origin is an isolated critical point, then $C_d(G, 0) \neq 0$.*

The following alternative obtained in Perera [106] gives a nontrivial critical point with a nontrivial critical group produced by a local linking.

Theorem 1.1.15 *If G has a local linking near the origin with $\dim N = d$, $H_d(G^b, G^a) = 0$ where $-\infty < a < 0 < b \leqslant +\infty$ are regular values, and G has only a finite number of critical points in G_a^b and satisfies $(C)_c$ for all $c \in [a, b] \cap \mathbb{R}$, then G has a critical point $u \neq 0$ with either*

$$
a < G(u) < 0, \qquad C_{d-1}(G, u) \neq 0
$$

or

$$
0 < G(u) < b, \qquad C_{d+1}(G, u) \neq 0.
$$

When G is bounded from below, taking $a < \inf G(E)$ and $b = +\infty$ gives the following three critical points theorem; see also Krasnosel'skii [62], Chang [28], Liu and Li [83], and Liu [82].

Corollary 1.1.16 *If G has a local linking near the origin with $\dim N = d \geqslant 2$, is bounded from below, has only a finite number of critical points, and satisfies (C), then G has a global minimizer $u_0 \neq 0$ with*

$$
G(u_0) < 0, \qquad C_q(G, u_0) = \delta_{q0}\, \mathcal{G}
$$

and a critical point $u \neq 0, u_0$ with either

$$
G(u) < 0, \qquad C_{d-1}(G, u) \neq 0
$$

or

$$
G(u) > 0, \qquad C_{d+1}(G, u) \neq 0.
$$

Proposition 1.1.14, Theorem 1.1.15, and Corollary 1.1.16 will be proved under a generalized notion of local linking in Section 1.9; see also Perera [107].

1.2 Compactness conditions

It is usually necessary to have some "compactness" when seeking critical points of a functional. The following condition was originally introduced by Palais and Smale [101].

Definition 1.2.1 G satisfies the Palais–Smale compactness condition at the level c, or $(PS)_c$ for short, if every sequence $(u_j) \subset E$ such that

$$G(u_j) \to c, \quad G'(u_j) \to 0,$$

called a $(PS)_c$ sequence, has a convergent subsequence; G satisfies (PS) if it satisfies $(PS)_c$ for every $c \in \mathbb{R}$, or equivalently, if every sequence such that

$$(G(u_j)) \text{ is bounded}, \quad G'(u_j) \to 0,$$

called a (PS) sequence, has a convergent subsequence.

The following weaker version was introduced by Cerami [25].

Definition 1.2.2 G satisfies the Cerami condition at the level c, or $(C)_c$ for short, if every sequence such that

$$G(u_j) \to c, \quad \left(1 + \|u_j\|\right) G'(u_j) \to 0,$$

called a $(C)_c$ sequence, has a convergent subsequence; G satisfies (C) if it satisfies $(C)_c$ for every c, or equivalently, if every sequence such that

$$(G(u_j)) \text{ is bounded}, \quad \left(1 + \|u_j\|\right) G'(u_j) \to 0,$$

called a (C) sequence, has a convergent subsequence.

This condition is weaker since a $(C)_c$ (resp. (C)) sequence is clearly a $(PS)_c$ (resp. (PS)) sequence also. Note that the limit of a $(PS)_c$ (resp. (PS)) sequence is in K^c (resp. K) since G is C^1. Since any sequence in K^c is a $(C)_c$ sequence, it follows that K^c is compact when $(C)_c$ holds.

1.3 Deformation lemmas

Deformation lemmas allow us to lower sublevel sets of a functional, away from its critical set, and are an essential tool for locating critical points. The main ingredient in their proofs is usually a suitable negative pseudo-gradient flow, a notion due to Palais [103].

Definition 1.3.1 A pseudo-gradient vector field for G on $\widehat{E}$ is a locally Lipschitz continuous mapping $V : \widehat{E} \to E$ satisfying

$$\|V(u)\| \leqslant \|G'(u)\|, \quad 2\left(G'(u), V(u)\right) \geqslant \|G'(u)\|^2 \quad \forall u \in \widehat{E}. \qquad (1.1)$$

Lemma 1.3.2 *There is a pseudo-gradient vector field V for G on $\widehat{E}$. When G is even, V may be chosen to be odd.*

Proof For each $u \in \widehat{E}$, there is a $w(u) \in E$ satisfying

$$\|w(u)\| < \|G'(u)\|, \quad 2\left(G'(u), w(u)\right) > \left(\|G'(u)\|\right)^2$$

by the definition of the norm in E'. Since G' is continuous, then

$$\|w(u)\| \leqslant \|G'(v)\|, \quad 2\left(G'(v), w(u)\right) \geqslant \left(\|G'(v)\|\right)^2 \quad \forall v \in N_u \qquad (1.2)$$

for some open neighborhood $N_u \subset \widehat{E}$ of u.

Since $\widehat{E}$ is a metric space and hence paracompact, the open covering $\left\{N_u\right\}_{u \in \widehat{E}}$ has a locally finite refinement, i.e. an open covering $\left\{N_\lambda\right\}_{\lambda \in \Lambda}$ of $\widehat{E}$ such that

(*i*) each $N_\lambda \subset N_{u_\lambda}$ for some $u_\lambda \in \widehat{E}$,
(*ii*) each $u \in \widehat{E}$ has a neighborhood U_u that intersects N_λ only for λ in some finite subset Λ_u of Λ.

(see, e.g., Kelley [61]). Let $\left\{\varphi_\lambda\right\}_{\lambda \in \Lambda}$ be a Lipschitz continuous partition of unity subordinate to $\left\{N_\lambda\right\}_{\lambda \in \Lambda}$, i.e.

(*i*) $\varphi_\lambda \in \mathrm{Lip}\left(\widehat{E}, [0, 1]\right)$ vanishes outside N_λ,
(*ii*) for each $u \in \widehat{E}$,

$$\sum_{\lambda \in \Lambda} \varphi_\lambda(u) = 1, \qquad (1.3)$$

where the sum is actually over a subset of Λ_u,

for example,

$$\varphi_\lambda(u) = \frac{\mathrm{dist}(u, \widehat{E} \backslash N_\lambda)}{\displaystyle\sum_{\lambda \in \Lambda} \mathrm{dist}(u, \widehat{E} \backslash N_\lambda)}.$$

Now

$$V(u) = \sum_{\lambda \in \Lambda} \varphi_\lambda(u)\, w(u_\lambda)$$

is Lipschitz in each U_u and satisfies (1.1) by (1.2) and (1.3).

When G is even, G' is odd and hence $-V(-u)$ is also a pseudo-gradient, and therefore so is the odd convex combination $\frac{1}{2}\left(V(u) - V(-u)\right)$. □

For $A \subset E$, let

$$N_\delta(A) = \{u \in E : \text{dist}(u, A) \leq \delta\}$$

be the δ-neighborhood of A. The following deformation lemma improves that of Cerami [25].

Lemma 1.3.3 (first deformation lemma) *If $c \in \mathbb{R}$, C is a bounded set containing K^c, $\delta, k > 0$, and G satisfies $(C)_c$, then there are an $\varepsilon_0 > 0$ and, for each $\varepsilon \in (0, \varepsilon_0)$, a map $\eta \in C(E \times [0, 1], E)$ satisfying*

(i) $\eta(\cdot, 0) = id_E$,
(ii) $\eta(\cdot, t)$ *is a homeomorphism of E for all $t \in [0, 1]$,*
(iii) $\eta(\cdot, t)$ *is the identity outside $A = G_{c-2\varepsilon}^{c+2\varepsilon}\backslash N_{\delta/3}(C)$ for all $t \in [0, 1]$,*
(iv) $\|\eta(u, t) - u\| \leq \left(1 + \|u\|\right)\delta/k \quad \forall(u, t) \in E \times [0, 1]$,
(v) $G(\eta(u, \cdot))$ *is nonincreasing for all $u \in E$,*
(vi) $\eta(G^{c+\varepsilon}\backslash N_\delta(C), 1) \subset G^{c-\varepsilon}$.

When G is even and C is symmetric, η may be chosen so that $\eta(\cdot, t)$ is odd for all $t \in [0, 1]$.

First we prove a lemma.

Lemma 1.3.4 *If $c \in \mathbb{R}$, N is an open neighborhood of K^c, $k > 0$, and G satisfies $(C)_c$, then there is an $\varepsilon_0 > 0$ such that*

$$\inf_{u \in G_{c-\varepsilon}^{c+\varepsilon}\backslash N} \left(1 + \|u\|\right)\left\|G'(u)\right\| \geq k\varepsilon \quad \forall \varepsilon \in (0, \varepsilon_0).$$

Proof If not, there are sequences $\varepsilon_j \searrow 0$ and $u_j \in G_{c-\varepsilon_j}^{c+\varepsilon_j}\backslash N$ such that

$$\left(1 + \|u_j\|\right)\left\|G'(u_j)\right\| < k\varepsilon_j.$$

Then $(u_j) \subset E\backslash N$ is a $(C)_c$ sequence and hence has a subsequence converging to some $u \in K^c\backslash N = \varnothing$, a contradiction. □

Proof of Lemma 1.3.3 Taking k larger if necessary, we may assume that

$$\left(1 + \|u\|\right)/k < 1/3 \quad \forall u \in N_\delta(C). \tag{1.4}$$

By Lemma 1.3.4, there is an $\varepsilon_0 > 0$ such that for each $\varepsilon \in (0, \varepsilon_0)$,

$$\left(1 + \|u\|\right)\left\|G'(u)\right\| \geq \frac{8\varepsilon}{\log(1 + \delta/k)} \quad \forall u \in A. \tag{1.5}$$

Let V be a pseudo-gradient vector field for G, $g \in \mathrm{Lip}_{\mathrm{loc}}(E, [0, 1])$ satisfy $g = 0$ outside A and $g = 1$ on $B = G_{c-\varepsilon}^{c+\varepsilon} \backslash N_{2\delta/3}(C)$, for example,

$$g(u) = \frac{\mathrm{dist}(u, E \backslash A)}{\mathrm{dist}(u, E \backslash A) + \mathrm{dist}(u, B)},$$

and $\eta(u, t)$, $0 \leqslant t < T(u) \leqslant \infty$ the maximal solution of

$$\dot{\eta} = -4\varepsilon \, g(\eta) \frac{V(\eta)}{\|V(\eta)\|^2}, \quad t > 0, \qquad \eta(u, 0) = u \in E. \tag{1.6}$$

For $0 \leqslant s < t < T(u)$,

$$\|\eta(u, t) - \eta(u, s)\| \leqslant 4\varepsilon \int_s^t \frac{g(\eta(u, \tau))}{\|V(\eta(u, \tau))\|} \, d\tau$$

$$\leqslant 8\varepsilon \int_s^t \frac{g(\eta(u, \tau))}{\|G'(\eta(u, \tau))\|} \, d\tau \qquad \text{by (1.1)}$$

$$\leqslant \log(1 + \delta/k) \int_s^t \left(1 + \|\eta(u, \tau)\| \right) d\tau \qquad \text{by (1.5)}$$

$$\leqslant \log(1 + \delta/k) \left[\int_s^t \|\eta(u, \tau) - \eta(u, s)\| \, d\tau \right.$$

$$\left. + \left(1 + \|\eta(u, s)\| \right) (t - s) \right],$$

and integrating gives

$$\|\eta(u, t) - \eta(u, s)\| \leqslant \left(1 + \|\eta(u, s)\| \right) \left((1 + \delta/k)^{t-s} - 1 \right). \tag{1.7}$$

Taking $s = 0$ we see that $\|\eta(u, \cdot)\|$ is bounded if $T(u) < \infty$, so $T(u) = \infty$ and (i)–(iv) follow.

By (1.6) and (1.1),

$$\frac{d}{dt} \left(G(\eta(u, t)) \right) = \left(G'(\eta), \dot{\eta} \right) = -4\varepsilon \, g(\eta) \frac{(G'(\eta), V(\eta))}{\|V(\eta)\|^2}$$

$$\leqslant -2\varepsilon \, g(\eta) \leqslant 0 \tag{1.8}$$

and hence (v) holds. To see that (vi) holds, let $u \in G^{c+\varepsilon} \backslash N_\delta(C)$ and suppose that $\eta(u, 1) \notin G^{c-\varepsilon}$. Then $\eta(u, t) \in G_{c-\varepsilon}^{c+\varepsilon}$ for all $t \in [0, 1]$, and we claim that $\eta(u, t) \notin N_{2\delta/3}(C)$. Otherwise there are $0 < s < t \leqslant 1$ such that

$$\mathrm{dist}(\eta(u, s), C) = \delta, \qquad 2\delta/3 < \mathrm{dist}(\eta(u, \tau), C) < \delta, \quad \tau \in (s, t),$$

$$\mathrm{dist}(\eta(u, t), C) = 2\delta/3.$$

But, then

$$\delta/3 \leqslant \|\eta(u, t) - \eta(u, s)\| \leqslant \left(1 + \|\eta(u, s)\|\right)\delta/k < \delta/3$$

by (1.7) and (1.4), a contradiction. Thus, $\eta(u, t) \in B$ and hence $g(\eta(u, t)) = 1$ for all $t \in [0, 1]$, so (1.8) gives

$$G(\eta(u, 1)) \leqslant G(u) - 2\varepsilon \leqslant c - \varepsilon,$$

a contradiction.

When G is even and C is symmetric, A and B are symmetric and hence g is even, so η may be chosen to be odd in u by choosing V to be odd. $\qquad\square$

If we assume $(PS)_c$ instead of $(C)_c$ in Lemma 1.3.3, then (iv) can be strengthened as follows.

Lemma 1.3.5 *If $c \in \mathbb{R}$, C is a set containing K^c, $\delta > 0$, $k > 3$, and G satisfies $(PS)_c$, then there are an $\varepsilon_0 > 0$ and, for each $\varepsilon \in (0, \varepsilon_0)$, a map $\eta \in C(E \times [0, 1], E)$ satisfying*

   (i) $\eta(\cdot, 0) = id_E$,
  (ii) $\eta(\cdot, t)$ *is a homeomorphism of E for all $t \in [0, 1]$,*
 (iii) $\eta(\cdot, t)$ *is the identity outside $A = G_{c-2\varepsilon}^{c+2\varepsilon} \backslash N_{\delta/3}(C)$ for all $t \in [0, 1]$,*
 (iv) $\|\eta(u, t) - u\| \leqslant \delta/k \quad \forall(u, t) \in E \times [0, 1]$,
  (v) $G(\eta(u, \cdot))$ *is nonincreasing for all $u \in E$,*
 (vi) $\eta(G^{c+\varepsilon} \backslash N_\delta(C), 1) \subset G^{c-\varepsilon}$.

When G is even and C is symmetric, η may be chosen so that $\eta(\cdot, t)$ is odd for all $t \in [0, 1]$.

First a lemma.

Lemma 1.3.6 *If $c \in \mathbb{R}$, N is an open neighborhood of K^c, $k > 0$, and G satisfies $(PS)_c$, then there is an $\varepsilon_0 > 0$ such that*

$$\inf_{u \in G_{c-\varepsilon}^{c+\varepsilon} \backslash N} \|G'(u)\| \geqslant k\varepsilon \quad \forall \varepsilon \in (0, \varepsilon_0).$$

Proof If not, there are sequences $\varepsilon_j \searrow 0$ and $u_j \in G_{c-\varepsilon_j}^{c+\varepsilon_j} \backslash N$ such that

$$\|G'(u_j)\| < k\varepsilon_j.$$

Then $(u_j) \subset E \backslash N$ is a $(PS)_c$ sequence and hence has a subsequence converging to some $u \in K^c \backslash N = \varnothing$, a contradiction. $\qquad\square$

Proof of Lemma 1.3.5 By Lemma 1.3.6, there is an $\varepsilon_0 > 0$ such that for each $\varepsilon \in (0, \varepsilon_0)$,

$$\|G'(u)\| \geqslant \frac{8k\,\varepsilon}{\delta} \quad \forall u \in A. \tag{1.9}$$

Let V be a pseudo-gradient vector field for G, $g \in \mathrm{Lip}_{\mathrm{loc}}(E, [0, 1])$ satisfy $g = 0$ outside A and $g = 1$ on $B = G_{c-\varepsilon}^{c+\varepsilon} \backslash N_{2\delta/3}(C)$, for example,

$$g(u) = \frac{\mathrm{dist}(u, E \backslash A)}{\mathrm{dist}(u, E \backslash A) + \mathrm{dist}(u, B)},$$

and $\eta(u, t)$, $0 \leqslant t < T(u) \leqslant \infty$ the maximal solution of

$$\dot{\eta} = -4\varepsilon\, g(\eta) \frac{V(\eta)}{\|V(\eta)\|^2}, \quad t > 0, \qquad \eta(u, 0) = u \in E. \tag{1.10}$$

Since

$$\|\eta(u, t) - u\| \leqslant 4\varepsilon \int_0^t \frac{g(\eta(u, \tau))}{\|V(\eta(u, \tau))\|}\, d\tau$$

$$\leqslant 8\varepsilon \int_0^t \frac{g(\eta(u, \tau))}{\|G'(\eta(u, \tau))\|}\, d\tau \qquad \text{by (1.1)}$$

$$\leqslant \frac{\delta t}{k} \qquad \text{by (1.9)},$$

$\|\eta(u, \cdot)\|$ is bounded if $T(u) < \infty$, so $T(u) = \infty$ and (i)–(iv) follow.

By (1.10) and (1.1),

$$\frac{d}{dt}\left(G(\eta(u, t))\right) = (G'(\eta), \dot{\eta}) = -4\varepsilon\, g(\eta) \frac{(G'(\eta), V(\eta))}{\|V(\eta)\|^2}$$

$$\leqslant -2\varepsilon\, g(\eta) \leqslant 0 \tag{1.11}$$

and hence (v) holds. To see that (vi) holds, let $u \in G^{c+\varepsilon} \backslash N_\delta(C)$ and suppose that $\eta(u, 1) \notin G^{c-\varepsilon}$. Then $\eta(u, t) \in G_{c-\varepsilon}^{c+\varepsilon}$ for all $t \in [0, 1]$, and $\eta(u, t) \notin N_{2\delta/3}(C)$ by (iv) since $k > 3$. Thus, $\eta(u, t) \in B$ and hence $g(\eta(u, t)) = 1$ for all $t \in [0, 1]$, so (1.11) gives

$$G(\eta(u, 1)) \leqslant G(u) - 2\varepsilon \leqslant c - \varepsilon,$$

a contradiction.

When G is even and C is symmetric, A and B are symmetric and hence g is even, so η may be chosen to be odd in u by choosing V to be odd. $\qquad \square$

First deformation lemma under the $(PS)_c$ condition is due to Palais [102]; see also Rabinowitz [126]. The proof under the $(C)_c$ condition was given by Cerami [25] and Bartolo *et al.* [13]; see also Perera *et al.* [113].

Lemma 1.3.3 provides a local deformation near a (possibly critical) level set of a functional. The following lemma shows that the homotopy type of sublevel sets can change only when crossing a critical level.

Lemma 1.3.7 (second deformation lemma) *If* $-\infty < a < b \leqslant +\infty$ *and* G *has only a finite number of critical points at the level* a, *has no critical values in* (a, b), *and satisfies* $(C)_c$ *for all* $c \in [a, b] \cap \mathbb{R}$, *then* G^a *is a strong deformation retract of* $G^b \backslash K^b$, *i.e. there is a map* $\eta \in C((G^b \backslash K^b) \times [0, 1], G^b \backslash K^b)$, *called a strong deformation retraction of* $G^b \backslash K^b$ *onto* G^a, *such that*

(i) $\eta(\cdot, 0) = id_{G^b \backslash K^b}$,
(ii) $\eta(\cdot, t)|_{G^a} = id_{G^a} \quad \forall t \in [0, 1]$,
(iii) $\eta(G^b \backslash K^b, 1) = G^a$.

Proof Let V be a pseudo-gradient vector field for G and $\zeta(u, t)$ the solution of

$$\dot{\zeta} = -\frac{V(\zeta)}{\|V(\zeta)\|^2}, \quad t > 0, \quad \zeta(u, 0) = u \in G^b \backslash (G^a \cup K^b) \subset \hat{E}. \quad (1.12)$$

Then

$$\frac{d}{dt}\left(G(\zeta(u, t))\right) = -\frac{(G'(\zeta), V(\zeta))}{\|V(\zeta)\|^2} \leqslant -\frac{1}{2}$$

by (1.1) and hence

$$G(\zeta(u, t)) \leqslant G(\zeta(u, s)) - \frac{1}{2}(t - s), \quad 0 \leqslant s < t. \quad (1.13)$$

Taking $s = 0$ and using $G(\zeta(u, 0)) = G(u) \leqslant b$ gives a $T(u) \in (0, 2(b - a)]$ such that $G(\zeta(u, t)) \searrow a$ as $t \nearrow T(u)$. We set $T(u) = 0$ and $\zeta(u, 0) = u$ for $u \in G^a$.

For $\delta > 0$, let

$$\mathcal{T}_\delta = \left\{t \in [0, T(u)) : \text{dist}(\zeta(u, t), K^a) \geqslant \delta\right\}.$$

Then

$$m_\delta := \inf_{t \in \mathcal{T}_\delta} \left(1 + \|\zeta(u, t)\|\right) \|G'(\zeta(u, t))\| > 0$$

by (C), so

$$\|\zeta(u, t) - \zeta(u, s)\| \leqslant \int_s^t \frac{d\tau}{\|V(\zeta(u, \tau))\|} \qquad \text{by (1.12)}$$

$$\leqslant 2 \int_s^t \frac{d\tau}{\|G'(\zeta(u, \tau))\|} \qquad \text{by (1.1)}$$

$$\leqslant \frac{2}{m_\delta} \int_s^t \left(1 + \|\zeta(u, \tau)\|\right) d\tau, \quad [s, t] \subset \mathcal{T}_\delta. \quad (1.14)$$

Case 1: $\zeta(u, \cdot)$ is bounded away from K^a. Then $\mathcal{T}_\delta = [0, T(u))$ for some $\delta > 0$ and $\|\zeta(u, \cdot)\|$ is bounded as in the proof of Lemma 1.3.3, so

$$\|\zeta(u, t) - \zeta(u, s)\| \leqslant C(t - s), \quad 0 \leqslant s < t < T(u) \qquad (1.15)$$

for some constant $C > 0$. Let $t_j \nearrow T(u)$. Taking $t = t_j$ and $s = t_k$ in (1.15) shows that $(\zeta(u, t_j))$ is a Cauchy sequence and hence converges to some $v \in G^{-1}(a) \backslash K^a$. Now taking $s = t_j$ shows that $\zeta(u, t) \to v$ as $t \nearrow T(u)$. We set $\zeta(u, T(u)) = v$ and note that T is continuous in u in this case.

Case 2: $\zeta(u, \cdot)$ is not bounded away from K^a. We claim that then $\zeta(u, t)$ converges to some $v \in K^a$ as $t \nearrow T(u)$. Since K^a is a finite set, otherwise there are a $v_0 \in K^a$, $\delta > 0$ such that the ball $B_{3\delta}(v_0)$ contains no other points of K^a, and sequences $s_j, t_j \nearrow T(u)$, $s_j < t_j$ such that

$$\|\zeta(u, s_j) - v_0\| = \delta, \qquad \delta < \|\zeta(u, \tau) - v_0\| < 2\delta, \quad \tau \in (s_j, t_j),$$

$$\|\zeta(u, t_j) - v_0\| = 2\delta.$$

But, then

$$\delta \leqslant \|\zeta(u, t_j) - \zeta(u, s_j)\| \leqslant C(t_j - s_j) \to 0$$

by (1.14), a contradiction.

We set $\zeta(u, T(u)) = v$ and claim that T is continuous in u in this case also. To see this, suppose that $u_j \to u$. We will show that

$$\underline{T} := \varliminf T(u_j) \geqslant T(u) \geqslant \varlimsup T(u_j) =: \overline{T}$$

and hence $T(u_j) \to T(u)$. If $\underline{T} < T(u)$, then passing to a subsequence, $\zeta(u_j, T(u_j)) \to \zeta(u, \underline{T})$ and hence

$$a = G(\zeta(u_j, T(u_j))) \to G(\zeta(u, \underline{T})) > a,$$

a contradiction. If $\overline{T} > T(u)$, then for a subsequence and any $t < T(u)$,

$$a = G(\zeta(u_j, T(u_j))) \leqslant G(\zeta(u_j, t)) - \frac{1}{2}(T(u_j) - t)$$

by (1.13), and passing to the limit gives

$$a \leqslant G(\zeta(u, t)) - \frac{1}{2}(\overline{T} - t) \to a - \frac{1}{2}(\overline{T} - T(u)) < a \quad \text{as } t \nearrow T(u),$$

again a contradiction.

We will show that ζ is continuous. Then

$$\eta(u, t) = \zeta(u, t\, T(u))$$

will be a strong deformation retraction of $G^b \backslash K^b$ onto G^a.

Case 1: $u \in G^b \backslash (G^a \cup K^b)$, $0 \leqslant t < T(u)$. Then ζ is continuous at (u, t) by standard ODE theory.

Case 2: $u \in G^b \backslash (G^a \cup K^b)$, $t = T(u)$. Suppose that $u_j \in G^b \backslash (G^a \cup K^b)$, $0 \leqslant t_j \leqslant T(u_j)$, $(u_j, t_j) \rightarrow (u, T(u))$, but $\zeta(u_j, t_j) \nrightarrow \zeta(u, T(u)) =:$ v. Then there is a $\delta > 0$ such that $(B_{3\delta}(v) \backslash \{v\}) \cap K^a = \varnothing$ and

$$\|\zeta(u_j, t_j) - v\| \geqslant 2\delta \qquad (1.16)$$

for a subsequence. Since $\zeta(u, s)$ converges to v as $s \nearrow T(u)$,

$$\|\zeta(u, s) - v\| \leqslant \delta/2 \qquad (1.17)$$

for all $s < T(u)$ sufficiently close to $T(u)$. For each such s,

$$\|\zeta(u_j, s) - \zeta(u, s)\| \leqslant \delta/2 \qquad (1.18)$$

for all sufficiently large j by Case 1. Taking a sequence $s_j \nearrow T(u)$ and combining (1.17) and (1.18) gives $t_j > s_j$ and

$$\|\zeta(u_j, s_j) - v\| \leqslant \delta \qquad (1.19)$$

for a further subsequence of (u_j, t_j). By (1.16) and (1.19), there are sequences $s'_j, t'_j \nearrow T(u)$, $s_j \leqslant s'_j < t'_j \leqslant t_j$ such that

$$\|\zeta(u_j, s'_j) - v\| = \delta, \qquad \delta < \|\zeta(\tau, u_j) - v\| < 2\delta, \quad \tau \in (s'_j, t'_j),$$

$$\|\zeta(u_j, t'_j) - v\| = 2\delta. \qquad (1.20)$$

Then

$$\delta \leqslant \|\zeta(u_j, t'_j) - \zeta(u_j, s'_j)\| \leqslant C(t'_j - s'_j) \rightarrow 0 \qquad (1.21)$$

by (1.14), a contradiction.

Case 3: $u \in G^{-1}(a)$, $t = 0$. Suppose that $u_j \in G^b \backslash K^b$, $0 \leqslant t_j \leqslant T(u_j)$, $(u_j, t_j) \rightarrow (u, 0)$, but $\zeta(u_j, t_j) \nrightarrow \zeta(u, 0) = u$. Then there is a $\delta > 0$ such that $(B_{3\delta}(u) \backslash \{u\}) \cap K^a = \varnothing$ and

$$\|\zeta(u_j, t_j) - u\| \geqslant 2\delta \qquad (1.22)$$

for a subsequence. Since $\zeta(u_j, 0) = u_j \rightarrow u$,

$$\|\zeta(u_j, 0) - u\| \leqslant \delta \qquad (1.23)$$

for sufficiently large j. By (1.22) and (1.23), $u_j \in G^b \backslash (G^a \cup K^b)$, $t_j > 0$, and there are sequences $0 \leqslant s'_j < t'_j \leqslant t_j$ for which (1.20), and hence also (1.21), holds.

Case 4: $u \in G^a \backslash G^{-1}(a)$, $t = 0$. Then $\zeta(u, 0) = u$. $\qquad \square$

Remark 1.3.8 Note that $G(\eta(u, t)) \leqslant G(u)$ for all $t \in [0, 1]$ by (1.13) and hence the restriction of η to $G^{-1}(-\infty, b) \times [0, 1]$ is a strong deformation retraction of $G^{-1}(-\infty, b)$ onto G^a.

The second deformation lemma under the $(PS)_c$ condition is due to Rothe [135], Chang [28], and Wang [157]. The proof under the $(C)_c$ condition was given by Bartsch and Li [14] and Perera and Schechter [119]; see also Perera *et al.* [113].

1.4 Critical groups

In Morse theory the local behavior of G near an isolated critical point u is described by the sequence of critical groups

$$C_q(G, u) = H_q(G^c \cap U, G^c \cap U \setminus \{u\}), \quad q \geqslant 0 \qquad (1.24)$$

where $c = G(u)$ is the corresponding critical value, U is a neighborhood of u, and H_* denotes singular homology with $\mathbb{Z}_2$-coefficients. They are independent of U, and hence well-defined, by the excision property. Critical groups help distinguish between different types of critical points and are extremely useful for obtaining multiple critical points of a functional (see, e.g., Chang [29]).

One of the consequences of the second deformation lemma is the following proposition relating the change in the topology of sublevel sets across a critical level to the critical groups of the critical points at that level.

Proposition 1.4.1 *If $-\infty < a < b \leqslant +\infty$ and G has only a finite number of critical points at the level $c \in (a, b)$, has no other critical values in $[a, b]$, and satisfies $(C)_{c'}$ for all $c' \in [a, b] \cap \mathbb{R}$, then*

$$H_q(G^b, G^a) \approx \bigoplus_{u \in K^c} C_q(G, u) \quad \forall q.$$

In particular,

$$\operatorname{rank} H_q(G^b, G^a) = \sum_{u \in K^c} \operatorname{rank} C_q(G, u) \quad \forall q.$$

Proof We have

$$H_q(G^b, G^a) \approx H_q(G^c, G^a) \approx H_q(G^c, G^c \setminus K^c) \qquad (1.25)$$

since G^c and G^a are strong deformation retracts of G^b and $G^c \setminus K^c$, respectively, by Lemma 1.3.7. Taking $\delta > 0$ so small that the balls $B_\delta(u)$, $u \in K^c$ are mutually disjoint and then excising $G^c \setminus \bigcup_{u \in K^c} B_\delta(u)$, we see that the last group in (1.25) is isomorphic to

$$\bigoplus_{u \in K^c} H_q(G^c \cap B_\delta(u), G^c \cap B_\delta(u) \setminus \{u\}) = \bigoplus_{u \in K^c} C_q(G, u). \qquad \square$$

For the change in the topology across multiple critical levels, we have the following.

Proposition 1.4.2 *If $-\infty < a < b \leqslant +\infty$ are regular values and G has only a finite number of critical points in G_a^b and satisfies* (C)$_c$ *for all $c \in [a, b] \cap \mathbb{R}$, then*

$$\operatorname{rank} H_q(G^b, G^a) \leqslant \sum_{u \in K_a^b} \operatorname{rank} C_q(G, u) \quad \forall q.$$

In particular, G has a critical point u with $a < G(u) < b$ and $C_q(G, u) \neq 0$ when $H_q(G^b, G^a) \neq 0$.

First we prove a lemma of a purely topological nature.

Lemma 1.4.3 *If $X_1 \subset \cdots \subset X_{k+1}$ are topological spaces, then*

$$\operatorname{rank} H_q(X_{k+1}, X_1) \leqslant \sum_{i=1}^{k} \operatorname{rank} H_q(X_{i+1}, X_i) \quad \forall q. \qquad (1.26)$$

Proof In the exact sequence

$$\cdots \longrightarrow H_q(X_k, X_1) \xrightarrow{\;i_*\;} H_q(X_{k+1}, X_1)$$
$$\xrightarrow{\;j_*\;} H_q(X_{k+1}, X_k) \longrightarrow \cdots$$

of the triple (X_{k+1}, X_k, X_1),

$$\operatorname{im} j_* \approx H_q(X_{k+1}, X_1) / \ker j_* = H_q(X_{k+1}, X_1) / \operatorname{im} i_*$$

and hence

$$\operatorname{rank} H_q(X_{k+1}, X_1) = \operatorname{rank} i_* + \operatorname{rank} j_*$$
$$\leqslant \operatorname{rank} H_q(X_k, X_1) + \operatorname{rank} H_q(X_{k+1}, X_k).$$

Since equality holds in (1.26) when $k = 1$, the conclusion now follows by induction on k. $\qquad \square$

Proof of Proposition 1.4.2 Let $c_1 < \cdots < c_k$ be the critical values in (a, b) and

$$a = a_1 < c_1 < a_2 < c_2 < \cdots < c_{k-1} < a_k < c_k < a_{k+1} = b.$$

Applying Lemma 1.4.3 with $X_i = G^{a_i}$ and using Proposition 1.4.1 gives

$$\operatorname{rank} H_q(G^b, G^a) \leqslant \sum_{i=1}^{k} \operatorname{rank} H_q(G^{i+1}, G^i)$$

$$= \sum_{i=1}^{k} \sum_{u \in K^{c_i}} \operatorname{rank} C_q(G, u)$$

$$= \sum_{u \in K_a^b} \operatorname{rank} C_q(G, u). \qquad \square$$

We refer to Chang and Ghoussoub [27] or Corvellec and Hantoute [32] for the proof of the following theorem, which shows that critical groups are invariant under homotopies that preserve the isolatedness of the critical point.

Theorem 1.4.4 *If G_t, $t \in [0, 1]$ is a family of C^1-functionals on E and u is a critical point of each G_t, with a closed neighborhood U such that*

(i) *G_t satisfies (PS) over U,*
(ii) *U contains no other critical points of G_t,*
(iii) *the map $[0, 1] \to C^1(U, \mathbb{R})$, $t \mapsto G_t$ is continuous,*

then

$$C_q(G_0, u) \approx C_q(G_1, u) \quad \forall q.$$

When the critical values are bounded from below and G satisfies (C), the global behavior of G can be described by the critical groups at infinity

$$C_q(G, \infty) = H_q(E, G^a), \quad q \geqslant 0 \qquad (1.27)$$

where a is less than all critical values. They are independent of a, and hence well-defined, by Lemma 1.3.7 and the homotopy invariance of the homology groups. Let

$$\delta_{qd} = \begin{cases} 1, & q = d \\ 0, & q \neq d \end{cases}$$

be the Kronecker delta and denote by $\widetilde{H}$ reduced homology. We have the following.

Proposition 1.4.5 *Assume that G satisfies (C).*

(i) *If G is bounded from below, then*

$$C_q(G, \infty) \approx \delta_{q0} \mathbb{Z}_2.$$

(*ii*) *If G is unbounded from below, then*

$$C_q(G, \infty) \approx \tilde{H}_{q-1}(G^a) \quad \forall q.$$

In particular, $C_0(G, \infty) = 0$.

Proof This follows from Lemma 1.4.6 below since E is contractible and $G^a = \emptyset$ if and only if G is bounded from below. $\square$

Lemma 1.4.6 *Let* (X, A) *be a pair with X contractible.*

(*i*) *If* $A = \emptyset$, *then*

$$H_q(X, A) \approx \delta_{q0} \mathbb{Z}_2.$$

(*ii*) *If* $A \neq \emptyset$, *then*

$$H_q(X, A) \approx \tilde{H}_{q-1}(A) \quad \forall q.$$

In particular, $H_0(X, A) = 0$.

Proof (*i*) Since $A = \emptyset$ and X is contractible,

$$H_q(X, A) = H_q(X) \approx \delta_{q0} \mathbb{Z}_2.$$

(*ii*) Since $A \neq \emptyset$ and the reduced groups are trivial in all dimensions for a contractible space, this follows from the exact sequence

$$\cdots \longrightarrow \tilde{H}_q(X) \longrightarrow H_q(X, A) \longrightarrow \tilde{H}_{q-1}(A)$$

$$\longrightarrow \tilde{H}_{q-1}(X) \longrightarrow \cdots . \qquad \square$$

When studying the existence and multiplicity of critical points of a functional we may often assume without loss of generality that there are only finitely many critical points. The following proposition relating the critical groups of G at infinity to those of its (finite) critical points is then immediate from Proposition 1.4.2 with $b = +\infty$.

Proposition 1.4.7 *If G satisfies* (C), *then*

$$\operatorname{rank} C_q(G, \infty) \leqslant \sum_{u \in K} \operatorname{rank} C_q(G, u) \quad \forall q.$$

In particular, G has a critical point u with $C_q(G, u) \neq 0$ *when* $C_q(G, \infty) \neq 0$.

1.5 Minimizers

In this section we give sufficient conditions for G to have a global minimizer and compute the critical groups at an isolated local minimizer.

Proposition 1.5.1 *If G is bounded from below and satisfies* $(C)_c$ *for $c = \inf G$, then G has a global minimizer.*

Proof If not, there are an $\varepsilon > 0$ and a map $\eta \in C(E \times [0, 1], E)$ satisfying $\eta(G^{c+\varepsilon}, 1) \subset G^{c-\varepsilon}$ by the first deformation lemma. Taking $u \in E$ with $G(u) \leqslant c + \varepsilon$, then we have $G(\eta(u, 1)) \leqslant c - \varepsilon < \inf G$, a contradiction. $\qquad \square$

Turning to critical groups, we have the next proposition.

Proposition 1.5.2 *If u is an isolated local minimizer of G, then*

$$C_q(G, u) \approx \delta_{q0} \, \mathbb{Z}_2.$$

Proof Let $c = G(u)$. Then for sufficiently small $r > 0$,

$$G(v) \geqslant c \quad \forall v \in B_r(u)$$

and there are no other critical points of G in $B_r(u)$. Since any $v \in B_r(u)$ with $G(v) = c$ is then a local minimizer,

$$G(v) > c, \quad v \in B_r(u) \backslash \{u\},$$

and hence

$$C_q(G, u) = H_q(G^c \cap B_r(u), G^c \cap B_r(u) \backslash \{u\}) = H_q(\{u\}, \varnothing) \approx \delta_{q0} \, \mathbb{Z}_2.$$

$$\square$$

Combining Propositions 1.4.5 (i), 1.5.1, and 1.5.2 gives the following corollary.

Corollary 1.5.3 *If G is bounded from below, satisfies (C), and has only a finite number of critical points, then $C_q(G, \infty) \approx \delta_{q0} \, \mathbb{Z}_2$ and G has a global minimizer u with $C_q(G, u) \approx \delta_{q0} \, \mathbb{Z}_2$.*

1.6 Nontrivial critical points

In many applications G has the trivial critical point $u = 0$ and we are interested in finding others. We assume that G has only a finite number of critical points.

When $u = 0$ is the only critical point of G, $C_q(G, 0) \approx C_q(G, \infty)$ for all q by Proposition 1.4.1 with $-\infty < a < 0 < b = +\infty$, so we have the following.

Proposition 1.6.1 *If* $C_q(G, 0) \not\approx C_q(G, \infty)$ *for some q and G satisfies* (C), *then G has a critical point* $u \neq 0$.

The following proposition is useful for obtaining a nontrivial critical point with a nontrivial critical group.

Proposition 1.6.2 *Assume that G satisfies* (C).

(i) *If* $C_q(G, 0) = 0$ *and* $C_q(G, \infty) \neq 0$ *for some q, then G has a critical point* $u \neq 0$ *with* $C_q(G, u) \neq 0$.

(ii) *If* $C_q(G, 0) \neq 0$ *and* $C_q(G, \infty) = 0$ *for some q, then G has a critical point* $u \neq 0$ *with either* $G(u) < 0$ *and* $C_{q-1}(G, u) \neq 0$, *or* $G(u) > 0$ *and* $C_{q+1}(G, u) \neq 0$.

First a purely topological lemma.

Lemma 1.6.3 *If* $X_1 \subset X_2 \subset X_3 \subset X_4$ *are topological spaces, then*

$$\text{rank } H_{q-1}(X_2, X_1) + \text{rank } H_{q+1}(X_4, X_3) \geq \text{rank } H_q(X_3, X_2)$$

$$- \text{rank } H_q(X_4, X_1) \quad \forall q.$$

Proof From the exact sequence

$$\cdots \longrightarrow H_q(X_3, X_1) \xrightarrow{j_*} H_q(X_3, X_2) \xrightarrow{\partial_*} H_{q-1}(X_2, X_1) \longrightarrow \cdots$$

of the triple (X_3, X_2, X_1), we have

$$\text{rank } H_q(X_3, X_1) \geq \text{rank } j_* = \text{nullity } \partial_* = \text{rank } H_q(X_3, X_2) - \text{rank } \partial_*$$

$$\geq \text{rank } H_q(X_3, X_2) - \text{rank } H_{q-1}(X_2, X_1),$$

and from the exact sequence

$$\cdots \longrightarrow H_{q+1}(X_4, X_3) \xrightarrow{\partial_*} H_q(X_3, X_1) \xrightarrow{i_*} H_q(X_4, X_1) \longrightarrow \cdots$$

of the triple (X_4, X_3, X_1),

$$\text{rank } H_q(X_3, X_1) = \text{rank } i_* + \text{nullity } i_* \leq \text{rank } H_q(X_4, X_1) + \text{rank } \partial_*$$

$$\leq \text{rank } H_q(X_4, X_1) + \text{rank } H_{q+1}(X_4, X_3),$$

so the conclusion follows. $\qquad\square$

Proof of Proposition 1.6.2 (i) By Proposition 1.4.7, G has a critical point u with $C_q(G, u) \neq 0$ since $C_q(G, \infty) \neq 0$, and $u \neq 0$ since $C_q(G, 0) = 0$.

(ii) Let $\varepsilon > 0$ be so small that zero is the only critical value in $[-\varepsilon, \varepsilon]$ and a be less than $-\varepsilon$ and all critical values. Since rank $H_q(G^\varepsilon, G^{-\varepsilon}) \geq$ rank $C_q(G, 0)$

by Proposition 1.4.1 and $H_q(E, G^a) = C_q(G, \infty)$, applying Lemma 1.6.3 to $G^a \subset G^{-\varepsilon} \subset G^{\varepsilon} \subset E$ gives

$$\operatorname{rank} H_{q-1}(G^{-\varepsilon}, G^a) + \operatorname{rank} H_{q+1}(E, G^{\varepsilon}) \geqslant \operatorname{rank} C_q(G, 0)$$

$$- \operatorname{rank} C_q(G, \infty) > 0.$$

Then either $H_{q-1}(G^{-\varepsilon}, G^a) \neq 0$, or $H_{q+1}(E, G^{\varepsilon}) \neq 0$, and the conclusion follows from Proposition 1.4.2. □

Remark 1.6.4 The alternative in Proposition 1.6.2 (*ii*) and Lemma 1.6.3 were proved by Perera [106, 107].

1.7 Mountain pass points

A critical point u of G with $C_1(G, u) \neq 0$ is called a mountain pass point. Since homology groups, and hence also critical groups, are trivial in negative dimensions, the special case $q = 0$ of Proposition 1.6.2 (*ii*) reduces to the following.

Corollary 1.7.1 *If* $C_0(G, 0) \neq 0$, $C_0(G, \infty) = 0$, *and G satisfies* (C), *then G has a mountain pass point* $u \neq 0$ *with* $G(u) > 0$.

This implies the well-known mountain pass lemma of Ambrosetti and Rabinowitz [9]. Indeed, if the origin is a local minimizer and G is unbounded from below, then $C_0(G, 0) \approx \mathbb{Z}_2$ by Proposition 1.5.2 and $C_0(G, \infty) = 0$ by Proposition 1.4.5 (*ii*), so Corollary 1.7.1 gives a positive mountain pass level.

1.8 Three critical points theorem

Another consequence of Proposition 1.6.2 (*ii*) is the following corollary.

Corollary 1.8.1 *If* $C_q(G, 0) \neq 0$ *for some* $q \geqslant 1$ *and G is bounded from below and satisfies* (C), *then G has a critical point* $u_1 \neq 0$. *If* $q \geqslant 2$, *then there is a second critical point* $u_2 \neq 0$.

Proof By Corollary 1.5.3, $C_q(G, \infty) = 0$ and G has a global minimizer u_1 with $C_q(G, u_1) = 0$. Since $C_q(G, 0) \neq 0$, $u_1 \neq 0$ and there is a critical point $u_2 \neq 0$ with either $C_{q-1}(G, u_2) \neq 0$ or $C_{q+1}(G, u_2) \neq 0$. When $q \geqslant 2$, $u_2 \neq u_1$ since $C_{q-1}(G, u_1) = C_{q+1}(G, u_1) = 0$. □

1.9 Generalized local linking

The notion of a generalized local linking is useful for obtaining nontrivial critical groups at zero and hence nontrivial critical points via Proposition 1.6.2 (ii).

Definition 1.9.1 We say that G has a generalized local linking near zero in dimension q, $1 \leqslant q < \infty$ if there are

(i) a direct sum decomposition $E = N \oplus M$, $u = v + w$ with $\dim N = q$,
(ii) $r > 0$ and a homeomorphism h from $C = \{u \in E : \|v\| \leqslant r, \|w\| \leqslant r\}$
onto a neighborhood U of zero such that $h(0) = 0$ and

$$G|_{h(C \cap N)} \leqslant 0 < G|_{h(C \cap M) \setminus \{0\}} . \tag{1.28}$$

Proposition 1.9.2 *If G has a generalized local linking near zero in dimension q, then $C_q(G, 0) \neq 0$.*

Proof We have $C_q(G, 0) = H_q(G^0 \cap U, G^0 \cap U \setminus \{0\})$ and the commutative diagram

$$
\begin{array}{ccc}
H_q(C \cap N, \partial C \cap N) & \xrightarrow[\approx]{h_*} & H_q(h(C \cap N), h(\partial C \cap N)) \\
\Big\downarrow{i_*} & & \Big\downarrow \\
 & H_q(G^0 \cap U, G^0 \cap U \setminus \{0\}) & \\
 & & \Big\downarrow \\
H_q(C, C \setminus M) & \xrightarrow[\approx]{h_*} & H_q(h(C), h(C \setminus M))
\end{array}
$$

where the vertical arrows are induced by inclusions. Since $C \cap N$ is a q-dimensional ball and $\partial C \cap N$ is its boundary, $H_q(C \cap N, \partial C \cap N) \approx \mathbb{Z}_2$, and since they are strong deformation retracts of C and $C \setminus M$, respectively, i_* is an isomorphism. $\qquad \square$

Remark 1.9.3 Definition 1.9.1 is a special case of the more general notion of a homological local linking introduced by Perera [107], which also yields a nontrivial critical group at zero. The strict inequality in (1.28) can be relaxed to less than or equal to; see Degiovanni *et al.* [46].

1.10 p-Laplacian

The p-Laplacian operator

$$\Delta_p u = \operatorname{div}\left(|\nabla u|^{p-2} \nabla u\right), \quad p \in (1, \infty)$$

arises in non-Newtonian fluid flows, turbulent filtration in porous media, plasticity theory, rheology, glacelogy, and in many other application areas; see, e.g., Esteban and Vázquez [50] and Padial *et al.* [100]. Problems involving the p-Laplacian have been studied extensively in the literature during the Past 50 years. In this section we present a result on nontrivial critical groups associated with the p-Laplacian obtained in Perera [110]; see also Dancer and Perera [41].

Consider the nonlinear eigenvalue problem

$$\begin{cases} -\Delta_p u = \lambda |u|^{p-2} u & \text{in } \Omega \\ \quad u = 0 & \text{on } \partial\Omega \end{cases} \qquad (1.29)$$

where Ω is a bounded domain in $\mathbb{R}^n$, $n \geqslant 1$ and $\Delta_p u = \text{div}\left(|\nabla u|^{p-2} \nabla u\right)$ is the p-Laplacian of u, $p \in (1, \infty)$. The set $\sigma(-\Delta_p)$ of eigenvalues is called the Dirichlet spectrum of $-\Delta_p$ on Ω. It is known that the first eigenvalue λ_1 is positive, simple, has an associated eigenfunction that is positive in Ω, and is isolated in the spectrum; see Anane [10] and Lindqvist [80, 81]. So the second eigenvalue $\lambda_2 = \inf \sigma(-\Delta_p) \cap (\lambda_1, \infty)$ is also defined; see Anane and Tsouli [11]. In the ODE case $n = 1$, where Ω is an interval, the spectrum consists of a sequence of simple eigenvalues $\lambda_k \nearrow \infty$ and the eigenfunction associated with λ_k has exactly $k - 1$ interior zeroes; see Cuesta [34] or Drábek [48]. In the semilinear PDE case $n \geqslant 2$, $p = 2$ also $\sigma(-\Delta)$ consists of a sequence of eigenvalues $\lambda_k \nearrow \infty$, but in the quasilinear PDE case $n \geqslant 2$, $p \neq 2$ a complete description of the spectrum is not available.

Eigenvalues of (1.29) are the critical values of the C^1-functional

$$I(u) = \int_\Omega |\nabla u|^p, \quad u \in S = \left\{u \in W = W_0^{1,p}(\Omega) : \|u\|_{L^p(\Omega)} = 1\right\},$$

which satisfies (PS). Denote by $\mathcal{A}$ the class of closed symmetric subsets of S and by

$$\gamma^+(A) = \sup\left\{k \geqslant 1 : \exists \text{ an odd continuous map } S^{k-1} \to A\right\},$$

$$\gamma^-(A) = \inf\left\{k \geqslant 1 : \exists \text{ an odd continuous map } A \to S^{k-1}\right\}$$

the cogenus and the genus of $A \in \mathcal{A}$, respectively, where S^{k-1} is the unit sphere in $\mathbb{R}^k$. Then

$$\lambda_k^\pm = \inf_{\substack{A \in \mathcal{A} \\ \gamma^\pm(A) \geqslant k}} \sup_{u \in A} I(u), \quad k \geqslant 1$$

are two increasing and unbounded sequences of eigenvalues (see Drábek and Robinson [49]), but, in general, it is not known whether either sequence is a complete list. The sequence $\left(\lambda_k^+\right)$ was introduced by Drábek and Robinson [49]; γ^- is also called the Krasnosel'skii genus [62].

Solutions of (1.29) are the critical points of the functional

$$I_\lambda(u) = \int_\Omega |\nabla u|^p - \lambda |u|^p, \quad u \in W_0^{1,p}(\Omega).$$

When $\lambda \notin \sigma(-\Delta_p)$, the origin is the only critical point of I_λ and hence the critical groups $C_q(I_\lambda, 0)$ are defined. Again we take the coefficient group to be $\mathbb{Z}_2$. The following theorem is our main result on them.

Theorem 1.10.1 ([110, Proposition 1.1]) *The spectrum of* $-\Delta_p$ *contains a sequence of eigenvalues* $\lambda_k \nearrow \infty$ *such that* $\lambda_k^- \leqslant \lambda_k \leqslant \lambda_k^+$ *and*

$$\lambda \in (\lambda_k, \lambda_{k+1}) \backslash \sigma(-\Delta_p) \implies C_k(I_\lambda, 0) \neq 0. \tag{1.30}$$

The eigenvalues λ_k are defined using the Yang index, whose definition and some properties we recall below. Various applications of this sequence of eigenvalues can be found in Perera [111, 112], Liu and Li [84], Perera and Szulkin [123], Cingolani and Degiovanni [31], Guo and Liu [58], Degiovanni and Lancelotti [44, 45], Tanaka [156], Fang and Liu [52], Medeiros and Perera [92], Motreanu and Perera [96], and Degiovanni *et al.* [46]. It is not known whether (1.30) holds for the sequences $(\lambda_k^\pm)$.

Yang [160] considered compact Hausdorff spaces with fixed-point-free continuous involutions and used the Čech homology theory, but for our purposes here it suffices to work with closed symmetric subsets of Banach spaces that do not contain the origin and singular homology groups.

Following [160], we first construct a special homology theory defined on the category of all pairs of closed symmetric subsets of Banach spaces that do not contain the origin and all continuous odd maps of such pairs. Let (X, A), $A \subset X$ be such a pair and $C(X, A)$ its singular chain complex with $\mathbb{Z}_2$-coefficients, and denote by $T_\#$ the chain map of $C(X, A)$ induced by the antipodal map $T(u) = -u$. We say that a q-chain c is symmetric if $T_\#(c) = c$, which holds if and only if $c = c' + T_\#(c')$ for some q-chain c'. The symmetric q-chains form a subgroup $C_q(X, A; T)$ of $C_q(X, A)$, and the boundary operator ∂_q maps $C_q(X, A; T)$ into $C_{q-1}(X, A; T)$, so these subgroups form a subcomplex $C(X, A; T)$. We denote by

$$Z_q(X, A; T) = \{c \in C_q(X, A; T) : \partial_q c = 0\},$$
$$B_q(X, A; T) = \{\partial_{q+1} c : c \in C_{q+1}(X, A; T)\},$$
$$H_q(X, A; T) = Z_q(X, A; T)/B_q(X, A; T)$$

the corresponding cycles, boundaries, and homology groups. A continuous odd map $f : (X, A) \to (Y, B)$ of pairs as above induces a chain map $f_\# : C(X, A; T) \to C(Y, B; T)$ and hence homomorphisms

$$f_* : H_q(X, A; T) \to H_q(Y, B; T).$$

For example,

$$H_q(S^k; T) = \begin{cases} \mathbb{Z}_2, & 0 \leqslant q \leqslant k \\ 0, & q \geqslant k+1 \end{cases}$$

(see [160, example 1.8]).

Let X be as above, and define homomorphisms $v : Z_q(X;T) \to \mathbb{Z}_2$ inductively by

$$v(z) = \begin{cases} \text{In}(c), & q = 0 \\ v(\partial c), & q > 0 \end{cases}$$

if $z = c + T_{\#}(c)$, where the index of a 0-chain $c = \sum_i n_i \sigma_i$ is defined by $\text{In}(c) = \sum_i n_i$. As in [160], v is well-defined and $v B_q(X;T) = 0$, so we can define the index homomorphism $v_* : H_q(X;T) \to \mathbb{Z}_2$ by $v_*([z]) = v(z)$. If F is a closed subset of X such that $F \cup T(F) = X$ and $A = F \cap T(F)$, then there is a homomorphism

$$\Delta : H_q(X;T) \to H_{q-1}(A;T)$$

such that $v_*(\Delta[z]) = v_*([z])$ (see [160, proposition 2.8]). Taking $F = X$ we see that if $v_* H_k(X;T) = \mathbb{Z}_2$, then $v_* H_q(X;T) = \mathbb{Z}_2$ for $0 \leqslant q \leqslant k$. We define the Yang index of X by

$$i(X) = \inf \left\{ k \geqslant -1 : v_* H_{k+1}(X;T) = 0 \right\},$$

taking $\inf \varnothing = \infty$. Clearly, $v_* H_0(X;T) = \mathbb{Z}_2$ if $X \neq \varnothing$, so $i(X) = -1$ if and only if $X = \varnothing$. For example, $i(S^k) = k$ (see [160, Example 3.4]).

Proposition 1.10.2 ([160, proposition 2.4]) *If $f : X \to Y$ is as above, then $v_*(f_*([z])) = v_*([z])$ for $[z] \in H_q(X;T)$, and hence $i(X) \leqslant i(Y)$. In particular, this inequality holds if $X \subset Y$.*

Thus, $k^+ - 1 \leqslant i(X) \leqslant k^- - 1$ if there are odd continuous maps $S^{k^+ - 1} \to X \to S^{k^- - 1}$, so

$$\gamma^+(X) \leqslant i(X) + 1 \leqslant \gamma^-(X). \tag{1.31}$$

Proposition 1.10.3 ([110, proposition 2.6]) *If $i(X) = k \geqslant 0$, then $\tilde{H}_k(X) \neq 0$.*

Proof We have

$$v_* H_q(X;T) = \begin{cases} \mathbb{Z}_2, & 0 \leqslant q \leqslant k \\ 0, & q \geqslant k+1. \end{cases}$$

We show that if $[z] \in H_k(X;T)$ is such that $\nu_*([z]) \neq 0$, then $[z] \neq 0$ in $\widetilde{H}_k(X)$. Arguing indirectly, assume that $z \in B_k(X)$, say, $z = \partial c$. Since $z \in B_k(X;T)$, $T_\#(z) = z$. Let $c' = c + T_\#(c)$. Then $c' \in Z_{k+1}(X;T)$ since $\partial c' = z + T_\#(z) = 2z = 0 \mod 2$, and $\nu_*([c']) = \nu(c') = \nu(\partial c) = \nu(z) \neq 0$, contradicting $\nu_* H_{k+1}(X;T) = 0$. □

Lemma 1.10.4 *We have*

$$C_q(I_\lambda, 0) \approx \widetilde{H}_{q-1}(I^\lambda) \quad \forall q.$$

Proof Taking $U = \{u \in W : \|u\|_{L^p(\Omega)} \leqslant 1\}$ in (1.24) gives

$$C_q(I_\lambda, 0) = H_q(I_\lambda^0 \cap U, I_\lambda^0 \cap U \setminus \{0\}).$$

Since I_λ is positive homogeneous, $I_\lambda^0 \cap U$ radially contracts to the origin via

$$(I_\lambda^0 \cap U) \times [0,1] \to I_\lambda^0 \cap U, \quad (u,t) \mapsto (1-t)u$$

and $I_\lambda^0 \cap U \setminus \{0\}$ deformation retracts onto $I_\lambda^0 \cap S$ via

$$(I_\lambda^0 \cap U \setminus \{0\}) \times [0,1] \to I_\lambda^0 \cap U \setminus \{0\}, \quad (u,t) \mapsto (1-t)u + t\,u/\|u\|_{L^p(\Omega)},$$

so it follows from the exact sequence of the pair $(I_\lambda^0 \cap U, I_\lambda^0 \cap U \setminus \{0\})$ that

$$H_q(I_\lambda^0 \cap U, I_\lambda^0 \cap U \setminus \{0\}) \approx \widetilde{H}_{q-1}(I_\lambda^0 \cap S).$$

Since $I_\lambda|_S = I - \lambda$, $I_\lambda^0 \cap S = I^\lambda$. □

We are now ready to prove Theorem 1.10.1.

Proof of Theorem 1.10.1 Set

$$\lambda_k = \inf_{\substack{A \in \mathcal{A} \\ i(A) \geqslant k-1}} \sup_{u \in A} I(u), \quad k \geqslant 1.$$

Then (λ_k) is an increasing sequence of critical values of I, and hence eigenvalues of $-\Delta_p$, by a standard deformation argument (see [110, proposition 3.1]). By (1.31), $\lambda_k^- \leqslant \lambda_k \leqslant \lambda_k^+$, in particular, $\lambda_k \to \infty$.

Let $\lambda \in (\lambda_k, \lambda_{k+1}) \setminus \sigma(-\Delta_p)$. By Lemma 1.10.4, $C_k(I_\lambda, 0) \approx \widetilde{H}_{k-1}(I^\lambda)$, and $I^\lambda \in \mathcal{A}$ since I is even. Since $\lambda > \lambda_k$, there is an $A \in \mathcal{A}$ with $i(A) \geqslant k-1$ such that $I \leqslant \lambda$ on A. Then $A \subset I^\lambda$ and hence $i(I^\lambda) \geqslant i(A) \geqslant k-1$ by Proposition 1.10.2. On the other hand, $i(I^\lambda) \leqslant k-1$ since $I \leqslant \lambda < \lambda_{k+1}$ on I^λ. So $i(I^\lambda) = k-1$ and hence $\widetilde{H}_{k-1}(I^\lambda) \neq 0$ by Proposition 1.10.3. □

2

Linking

2.1 Introduction

The purpose of this chapter is to introduce the reader to linking methods used in variational problems. General references are Rabinowitz [129], Struwe [153], Chang [29], Benci [17], and Schechter [143]; see also Perera *et al.* [113]. Our focus here will be the notion of homological linking, in particular, we will obtain pairs of critical points with nontrivial critical groups. Nonstandard geometries without a finite-dimensional closed loop will also be considered.

2.2 Minimax principle

Minimax principle originated in the work of Ljusternik and Schnirelmann [85] and is a useful tool for finding critical points of a functional. By Lemma 1.3.3, if c is a regular value, G satisfies $(C)_c$, and $\varepsilon > 0$ is sufficiently small, there is a map $\eta \in C(E \times [0, 1], E)$ such that, writing $\eta(u, t) = \eta(t) u$,

(i) $\eta(0) = id_E$,
(ii) $\eta(t)$ is a homeomorphism of E for all $t \in [0, 1]$, and the mapping $\eta^{-1} : E \times [0, 1] \to E$, $(u, t) \mapsto \eta(t)^{-1} u$ is continuous,
(iii) $\eta(t)$ is the identity on $E \backslash G_{c-2\varepsilon}^{c+2\varepsilon}$ for all $t \in [0, 1]$,
(iv) $G(\eta(\cdot) u)$ is nonincreasing for each $u \in E$,
(v) $\eta(1) G^{c+\varepsilon} \subset G^{c-\varepsilon}$,

and

$$\|\eta(t) u - u\| \leqslant \left(1 + \|u\|\right)/2 \quad \forall (u, t) \in E \times [0, 1].$$

Then

$$\|\eta(t)\,u\| \leqslant \left(1 + 3\,\|u\|\right)/2, \quad \|\eta^{-1}(t)\,u\| \leqslant 1 + 2\,\|u\| \quad \forall (u,t) \in E \times [0,1],$$

so we have

(*vi*) for each bounded subset A of E,

$$\sup_{(u,t) \in A \times [0,1]} \left(\|\eta(t)\,u\| + \|\eta^{-1}(t)\,u\| \right) < \infty.$$

Denote by $\mathcal{D}_{c,\varepsilon}$ the set of all maps $\eta \in C(E \times [0,1], E)$ satisfying (*i*)–(*vi*). We say that a family $\mathcal{F}$ of subsets of E is invariant under $\mathcal{D}_{c,\varepsilon}$ if

$$M \in \mathcal{F}, \ \eta \in \mathcal{D}_{c,\varepsilon} \implies \eta(1)\,M \in \mathcal{F}.$$

Proposition 2.2.1 (minimax principle) *If $\mathcal{F}$ is a family of subsets of E,*

$$c := \inf_{M \in \mathcal{F}} \sup_{u \in M} G(u)$$

is finite, $\mathcal{F}$ is invariant under $\mathcal{D}_{c,\varepsilon}$ for all sufficiently small $\varepsilon > 0$, and G satisfies $(C)_c$*, then c is a critical value of G.*

Proof If not, taking $\varepsilon > 0$ sufficiently small, $M \in \mathcal{F}$ with $\sup G(M) \leqslant c + \varepsilon$, and $\eta \in \mathcal{D}_{c,\varepsilon}$, we have $\eta(1)\,M \in \mathcal{F}$ and $\sup G(\eta(1)\,M) \leqslant c - \varepsilon$ by (*v*), contradicting the definition of c. □

Some references for Proposition 2.2.1 are Palais [103], Nirenberg [99], Rabinowitz [129], Schechter and Tintarev [148], Ghoussoub [56], and Schechter [142, 143]. Minimax methods were introduced in Morse theory by Marino and Prodi [89]; see also Liu [82].

2.3 Homotopical linking

The notion of homotopical linking is useful for obtaining critical points via the minimax principle.

Definition 2.3.1 Let D be a subset of E homeomorphic to the unit disk in $\mathbb{R}^n$ for some $n \geqslant 1$, A the relative boundary of D, B a nonempty subset of E disjoint from A, and

$$\Psi = \left\{ h \in C(D, E) : h|_A = id_A \right\}.$$

We say that A *homotopically links* B if

$$h(D) \cap B \neq \varnothing \quad \forall h \in \Psi. \tag{2.1}$$

Some standard examples are the following. Proofs are postponed till the next section, where we will show that these sets link homologically and that homological linking implies homotopical linking.

Example 2.3.2 If $u_0 \in E$, U is a bounded neighborhood of u_0, and $u_1 \notin \overline{U}$, then $A = \{u_0, u_1\}$ homotopically links $B = \partial U$.

Example 2.3.3 If $E = N \oplus M$, $u = v + w$ is a direct sum decomposition with N nontrivial and finite dimensional, then $A = \{v \in N : \|v\| = R\}$ homotopically links $B = M$ for any $R > 0$. In particular, if E is finite dimensional, then $A = \{u \in E : \|u\| = R\}$ homotopically links $B = \{0\}$.

Example 2.3.4 If M and N are as above and $w_0 \in M$ with $\|w_0\| = 1$, then $A = \{v \in N : \|v\| \leqslant R\} \cup \{u = v + tw_0 : v \in N, \ t \geqslant 0, \ \|u\| = R\}$ homotopically links $B = \{w \in M : \|w\| = r\}$ for any $0 < r < R$.

Theorem 2.3.5 *If A homotopically links the closed set B,*

$$c := \inf_{h \in \Psi} \max_{u \in D} G(h(u)),$$

$a := \sup G(A) \leqslant \inf G(B) =: b$, and G satisfies $(C)_c$, then $c \geqslant b$ and is a critical value of G. If $c = b$, then G has a critical point with critical value c on B.

Proof By (2.1), $c \geqslant b$. First suppose that $c > b$, and let $2\varepsilon < c - a$. Then for any $\eta \in \mathcal{D}_{c,\varepsilon}$, $\eta(t)$ is the identity on A for all $t \in [0, 1]$ by *(iii)* in the definition of $\mathcal{D}_{c,\varepsilon}$, so $\eta(1) \circ h \in \Psi$ whenever $h \in \Psi$. So $\mathcal{F} = \{h(D) : h \in \Psi\}$ is invariant under $\mathcal{D}_{c,\varepsilon}$, and hence c is a critical value of G by Proposition 2.2.1.

Now suppose that $c = b$ and $K^c \cap B = \varnothing$. Since A is compact and so is K^c by $(C)_c$, $C = A \cup K^c$ is compact. Since B is closed, then $\mathrm{dist}(C, B) > 0$. Applying Lemma 1.3.3 to $-G$ with this C and $\delta < \mathrm{dist}(C, B)$ gives an $\varepsilon > 0$ and a homeomorphism ζ of E such that ζ is the identity outside $G^{c+2\varepsilon}_{c-2\varepsilon} \backslash N_{\delta/3}(C)$ and

$$\zeta(G_{c-\varepsilon} \backslash N_\delta(C)) \subset G_{c+\varepsilon},$$

in particular, ζ is the identity on A and $G \geqslant c + \varepsilon$ on $\zeta(B)$. Then taking a $h \in \Psi$ with $G < c + \varepsilon$ on $h(D)$, we have $\tilde{h} := \zeta^{-1} \circ h \in \Psi$ and hence

$$h(D) \cap \zeta(B) = \zeta(\tilde{h}(D) \cap B) \neq \varnothing$$

by (2.1), a contradiction.   □

Many authors have contributed to this result. The special cases that correspond to Examples 2.3.2, 2.3.3, and 2.3.4 are the well-known mountain pass lemma of Ambrosetti and Rabinowitz [9] and the saddle point and linking theorems of Rabinowitz [127, 128], respectively. See also Ahmad *et al.* [2], Castro and Lazer [24], Benci and Rabinowitz [18], Ni [97], Chang [29], Qi [125], and Ghoussoub [55]. Morse index estimates for a critical point produced by a homotopical linking have been obtained by Lazer and Solimini [67], Solimini [152], Ghoussoub [55], Ramos and Sanchez [130], and others. However, the notion of homological linking introduced by Benci [15, 16] and Liu [82] is better suited for obtaining critical points with nontrivial critical groups.

2.4 Homological linking

The notion of homological linking is useful for obtaining pairs of sublevel sets with nontrivial relative homology and hence critical points with nontrivial critical groups via Proposition 1.4.2.

Definition 2.4.1 Let A and B be disjoint nonempty subsets of E. We say that A homologically links B in dimension $q < \infty$ if the homomorphism

$$i_* : \tilde{H}_q(A) \to \tilde{H}_q(E \backslash B),$$

induced by the inclusion $i : A \subset E \backslash B$, is nontrivial.

Note that $q \leqslant \dim E - 1$ when E is finite dimensional. The following three propositions show that the sets in the examples of homotopical linking given in the previous section link homologically also.

Proposition 2.4.2 If $u_0 \in E$, U is a bounded neighborhood of u_0, and $u_1 \notin \overline{U}$, then $A = \{u_0, u_1\}$ homologically links $B = \partial U$ in dimension $q = 0$.

Proof Follows since u_0 and u_1 are in different path components of $E \backslash B$. $\square$

Proposition 2.4.3 If $E = N \oplus M$, $u = v + w$ is a direct sum decomposition with $1 \leqslant d := \dim N < \infty$, then $A = \{v \in N : \|v\| = R\}$ homologically links $B = M$ in dimension $q = d - 1$ for any $R > 0$. In particular, if $d := \dim E < \infty$, then $A = \{u \in E : \|u\| = R\}$ homologically links $B = \{0\}$ in dimension $q = d - 1$.

Proof The map

$$(E \backslash B) \times [0, 1] \to E \backslash B, \quad (u, t) \mapsto (1 - t)u + t \frac{Rv}{\|v\|}$$

is a strong deformation retraction of $E \backslash B$ onto A and hence i_* is an isomorphism, so

$$\operatorname{rank} i_* = \operatorname{rank} \widetilde{H}_{d-1}(A) = 1. \qquad \square$$

Proposition 2.4.4 *If $E = N \oplus M$, $u = v + w$ is a direct sum decomposition with $1 \leqslant d := \dim N < \infty$ and $w_0 \in M$ with $\|w_0\| = 1$, then $A = \{v \in N : \|v\| \leqslant R\} \cup \{u = v + tw_0 : v \in N, t \geqslant 0, \|u\| = R\}$ homologically links $B = \{w \in M : \|w\| = r\}$ in dimension $q = d$ for any $0 < r < R$.*

Proof Let

$$
\begin{aligned}
A_0 &= \{v \in N : \|v\| = R\}, \qquad A_1 = \{v \in N : \|v\| \leqslant R\}, \\
A_2 &= \{u = v + tw_0 : v \in N, t \geqslant 0, \|u\| = R\}, \\
A' &= \{u = v + tw_0 : v \in N, t \geqslant 0, \|u\| \leqslant R\}, \\
B' &= \{w \in M : \|w\| \leqslant r\}, \qquad B'' = E \backslash \{w \in M : \|w\| \geqslant r\},
\end{aligned}
$$

and consider the commutative diagram

$$
\begin{array}{ccccc}
H_d(A_1, A_0) & \xrightarrow{\approx} & \widetilde{H}_{d-1}(A_0) & \longrightarrow & \widetilde{H}_{d-1}(A_1) \\
{\scriptstyle k_*}\downarrow & & {\scriptstyle j_*}\downarrow & & \downarrow \\
H_d(B'', E\backslash M) & \longrightarrow & \widetilde{H}_{d-1}(E\backslash M) & \longrightarrow & \widetilde{H}_{d-1}(B'')
\end{array}
$$

where the rows come from the exact sequences of the pairs $(A_1, A_0) \subset (B'', E\backslash M)$. Since $j_* \neq 0$ by Proposition 2.4.3 and A_1 is contractible, $k_* \neq 0$.

Now consider the commutative diagram

$$
\begin{array}{ccc}
H_d(A_1, A_0) & \longrightarrow & H_d(A, A_2) \\
{\scriptstyle k_*}\downarrow & & \downarrow{\scriptstyle l_*} \\
H_d(B'', E\backslash M) & \xrightarrow{\approx} & H_d(E\backslash B, E\backslash B')
\end{array}
$$

induced by inclusions. The bottom arrow is an isomorphism by the excision property since $\{w \in M : \|w\| > r\}$ is a closed subset of $E\backslash B$ contained in the open subset $E\backslash B'$. Since $k_* \neq 0$, then $l_* \neq 0$.

Next consider the commutative diagram

$$
\begin{array}{ccccc}
H_{d+1}(A', A) & \xrightarrow{\approx} & H_d(A, A_2) & \longrightarrow & H_d(A', A_2) \\
{\scriptstyle m_*}\downarrow & & \downarrow{\scriptstyle l_*} & & \downarrow \\
H_{d+1}(E, E\backslash B) & \longrightarrow & H_d(E\backslash B, E\backslash B') & \longrightarrow & H_d(E, E\backslash B')
\end{array}
$$

where the rows come from the exact sequences of the triples $(A', A, A_2) \subset (E, E\backslash B, E\backslash B')$. Since $l_* \neq 0$ and A' and A_2 are contractible, $m_* \neq 0$.

Finally consider the commutative diagram

$$\begin{array}{ccccc}
H_{d+1}(A', A) & \longrightarrow & H_d(A) & \longrightarrow & H_d(A') \\
{\scriptstyle m_*}\downarrow & & {\scriptstyle i_*}\downarrow & & \downarrow \\
H_{d+1}(E, E\backslash B) & \xrightarrow{\approx} & H_d(E\backslash B) & \longrightarrow & H_d(E)
\end{array}$$

where the rows come from the exact sequences of the pairs $(A', A) \subset (E, E\backslash B)$. Since $m_* \neq 0$ and E is contractible, $i_* \neq 0$. $\qquad\square$

The following proposition shows that homological linking is invariant under homeomorphisms of the space.

Proposition 2.4.5 *If A homologically links B in dimension q and h is a homeomorphism of E, then $h(A)$ homologically links $h(B)$ in dimension q.*

Proof Follows from the commutative diagram

$$\begin{array}{ccc}
\tilde{H}_q(A) & \xrightarrow{i_*} & \tilde{H}_q(E\backslash B) \\
{\scriptstyle h_*}\downarrow {\scriptstyle \approx} & & {\scriptstyle h_*}\downarrow {\scriptstyle \approx} \\
\tilde{H}_q(h(A)) & \xrightarrow{j_*} & \tilde{H}_q(h(E\backslash B))
\end{array}$$

where $j : h(A) \subset h(E\backslash B)$. $\qquad\square$

The following proposition shows that homological linking implies homotopical linking.

Proposition 2.4.6 *Let A and B be as in the Definition 2.3.1. If A homologically links B in dimension $n - 1$, then A homotopically links B.*

Proof Since $i_* \neq 0$ and the homology class $[id_A]$ generates $\tilde{H}_{n-1}(A)$, $i_* [id_A] \neq 0$ in $\tilde{H}_{n-1}(E\backslash B)$. So there is no singular n-chain of $E\backslash B$ with boundary id_A, in particular, there is no map $h \in C(D, E\backslash B)$ with $h|_A = id_A$. $\qquad\square$

Proposition 2.4.7 *If A homologically links B in dimension q and $G|_A \leqslant a < G|_B$, then $H_{q+1}(E, G^a) \neq 0$. In particular, $C_{q+1}(G, \infty) \neq 0$ when a is less than all critical values and G satisfies* (C).

Proof Since $i_* \neq 0$ in the commutative diagram

$$\begin{array}{ccc}
\tilde{H}_q(A) & \longrightarrow & \tilde{H}_q(G^a) \\
& {\scriptstyle i_*}\searrow & \downarrow \\
& & \tilde{H}_q(E\backslash B)
\end{array}$$

induced by the inclusions $A \subset G^a \subset E \backslash B$, $\tilde{H}_q(G^a) \neq 0$. Since E is contractible, $H_{q+1}(E, G^a) \approx \tilde{H}_q(G^a)$ by Lemma 1.4.6 (*ii*). □

Combining Propositions 2.4.7 and 1.4.2 gives the following theorem.

Theorem 2.4.8 *If A homologically links B in dimension q, $G|_A \leqslant a < G|_B$ where a is a regular value, and G has only a finite number of critical points in G_a and satisfies* $(\text{C})_c$ *for all $c \geqslant a$, then G has a critical point u with*

$$G(u) > a, \quad C_{q+1}(G, u) \neq 0.$$

2.5 Schechter and Tintarev's notion of linking

Schechter and Tintarev [148] introduced a different notion of linking that yields pairs of critical points, and the following refined version of their definition was given in Schechter [142]. Denote by Φ the set of all maps $\Gamma \in C(E \times [0, 1], E)$ such that, writing $\Gamma(u, t) = \Gamma(t) u$,

(*i*) $\Gamma(0) = id_E$,
(*ii*) $\Gamma(t)$ is a homeomorphism of E for all $t \in [0, 1)$, and the mapping $\Gamma^{-1} : E \times [0, 1) \to E$, $(u, t) \mapsto \Gamma(t)^{-1} u$ is continuous,
(*iii*) $\Gamma(1) E$ is a single point in E, and $\Gamma(t) u \to \Gamma(1) E$ as $t \to 1$, uniformly on bounded subsets of E,
(*iv*) for each bounded subset A of E and $t_0 \in [0, 1)$,

$$\sup_{(u,t) \in A \times [0, t_0]} \left(\|\Gamma(t) u\| + \|\Gamma^{-1}(t) u\| \right) < \infty.$$

Remark 2.5.1 Note that if $\Gamma \in \Phi$ and A is a bounded subset of E, $\Gamma(A \times (t_0, 1])$ is bounded for $t_0 < 1$ sufficiently close to 1 by (*iii*), and hence $\Gamma(A \times [0, 1])$ is bounded by (*iv*).

Definition 2.5.2 Let A and B be disjoint nonempty subsets of E. We say that A *links* B if

$$\Gamma(A \times (0, 1]) \cap B \neq \varnothing \quad \forall \Gamma \in \Phi. \tag{2.2}$$

Note that if $\Gamma \in \Phi$ and the map $\eta \in C(E \times [0, 1], E)$ is such that, writing $\eta(u, t) = \eta(t) u$,

(*i*) $\eta(0) = id_E$,
(*ii*) $\eta(t)$ is a homeomorphism of E for all $t \in [0, 1]$, and the mapping $\eta^{-1} : E \times [0, 1] \to E$, $(u, t) \mapsto \eta(t)^{-1} u$ is continuous,

(*iii*) for each bounded subset A of E,

$$\sup_{(u,t)\in A\times[0,1]} \left(\|\eta(t)\,u\| + \|\eta^{-1}(t)\,u\| \right) < \infty,$$

then the map

$$\eta \cdot \Gamma : E \times [0,1] \to E, \quad (u,t) \mapsto \begin{cases} \eta(2t)\,u, & t \in [0, 1/2] \\ \eta(1)\,\Gamma(2t-1)\,u, & t \in (1/2, 1] \end{cases}$$

is in Φ. In this sense, Φ is invariant under the family of maps $\mathcal{D}_{c,\varepsilon}$ defined in Section 2.2. Analogous to Theorem 2.3.5, we have the following.

Theorem 2.5.3 *If A is bounded and links the closed set B, $\text{dist}(A, B) > 0$,*

$$c := \inf_{\Gamma \in \Phi} \sup_{(u,t)\in A\times[0,1]} G(\Gamma(t)\,u)$$

is finite, $a := \sup G(A) \leqslant \inf G(B) =: b$, and G satisfies $(C)_c$, then $c \geqslant b$ and is a critical value of G. If $c = b$, then G has a critical point with critical value c on B.

Proof By (2.2), $c \geqslant b$. First suppose that $c > b$, and let $2\varepsilon < c - a$. Then for any $\eta \in \mathcal{D}_{c,\varepsilon}$, $\eta(t)$ is the identity on A for all $t \in [0, 1]$ by (*iii*) in the definition of $\mathcal{D}_{c,\varepsilon}$, so $\eta(1)\,\Gamma(A \times [0, 1]) = \eta \cdot \Gamma\,(A \times [0, 1])$ for all $\Gamma \in \Phi$. So $\mathcal{F} = \{\Gamma(A \times [0, 1]) : \Gamma \in \Phi\}$ is invariant under $\mathcal{D}_{c,\varepsilon}$, and hence c is a critical value of G by Proposition 2.2.1.

Now suppose that $c = b$ and $K^c \cap B = \varnothing$. Since K^c is compact by $(C)_c$ and B is closed, $\text{dist}(K^c, B) > 0$. Applying Lemma 1.3.3 to $-G$ with $C = A \cup K^c$ and $\delta < \text{dist}(C, B)$ gives an $\varepsilon > 0$ and a map $\zeta \in C(E \times [0, 1], E)$ such that, writing $\zeta(u, t) = \zeta(t)\,u$,

(*i*) $\zeta(0) = id_E$,
(*ii*) $\zeta(t)$ is a homeomorphism of E for all $t \in [0, 1]$, and the mapping $\zeta^{-1} : E \times [0, 1] \to E$, $(u, t) \mapsto \zeta(t)^{-1}\,u$ is continuous,
(*iii*) $\zeta(t)$ is the identity outside $G_{c-2\varepsilon}^{c+2\varepsilon} \backslash N_{\delta/3}(C)$ for all $t \in [0, 1]$,
(*iv*) $G(\zeta(\cdot)\,u)$ is nondecreasing for each $u \in E$,
(*v*) $\zeta(1)\,(G_{c-\varepsilon}\backslash N_\delta(C)) \subset G_{c+\varepsilon}$,
(*vi*) for each bounded subset Q of E,

$$\sup_{(u,t)\in Q\times[0,1]} \left(\|\zeta(t)\,u\| + \|\zeta^{-1}(t)\,u\| \right) < \infty.$$

In particular, $\zeta(t)$ is the identity on A for all $t \in [0, 1]$ and $G \geqslant c + \varepsilon$ on $\zeta(1)\,B$. Then taking a $\Gamma \in \Phi$ with $G < c + \varepsilon$ on $\Gamma(A \times [0, 1])$, we have

$\tilde{\Gamma} := \zeta^{-1} \cdot \Gamma \in \Phi$ and $\tilde{\Gamma}(A \times [0, 1]) = \zeta^{-1}(1)\,\Gamma(A \times [0, 1])$, so

$$\Gamma(A \times [0, 1]) \cap \zeta(1)\,B = \zeta(1)\,(\tilde{\Gamma}(A \times [0, 1]) \cap B) \neq \varnothing$$

by (2.2), a contradiction. $\qquad\square$

The following proposition shows that homotopical linking implies linking in this sense.

Proposition 2.5.4 *Let A and B be as in the Definition 2.3.1. If A homotopically links B, then A links B.*

Proof If $\Gamma \in \Phi$ and φ is a homeomorphism of D onto the unit disk in $\mathbb{R}^n$, then the map

$$h : D \to E, \quad u \mapsto \begin{cases} \Gamma(1 - \|\varphi(u)\|)\,\varphi^{-1}(\varphi(u)/\|\varphi(u)\|), & \varphi(u) \neq 0 \\[2mm] \Gamma(1)\,E, & \varphi(u) = 0 \end{cases}$$

is in Ψ and $h(D) = \Gamma(A \times [0, 1])$, so (2.1) implies (2.2). $\qquad\square$

Combining Propositions 2.4.6 and 2.5.4 gives

$$\begin{array}{ccccc} \text{homological} & & \text{homotopical} & & \text{Schechter–Tintarev} \\ \text{linking} & \Longrightarrow & \text{linking} & \Longrightarrow & \text{linking.} \end{array}$$

In particular, A links B in Examples 2.3.2 – 2.3.4. Note that when E is infinite dimensional the unit sphere in E is contractible and therefore does not link the origin homologically or homotopically. However, it does so according to Schechter and Tintarev's definition of linking, as the following proposition shows.

Proposition 2.5.5 *If U is a bounded neighborhood of $u_0 \in E$, then $A = \partial U$ links $B = \{u_0\}$.*

Proof Suppose not, say,

$$u_0 \notin \Gamma_0(\partial U \times [0, 1]) \tag{2.3}$$

where $\Gamma_0 \in \Phi$. Then $\Gamma_0(1)\,E \neq u_0$ and hence

$$\text{dist}(\Gamma_0(t_0)\,\overline{U}, u_0) > 0$$

for $t_0 < 1$ sufficiently close to 1 by (*iii*) in the definition of Φ, so $\Gamma_0^{-1}(t_0)\,u_0 \notin \overline{U}$. Thus, the path $\gamma(t) = \Gamma_0^{-1}(t)\,u_0$, $t \in [0, t_0]$ satisfies $\gamma(0) \in U$ and $\gamma(t_0) \notin \overline{U}$. But then $\gamma(t_1) \in \partial U$ for some $t_1 \in (0, t_0)$, so $u_0 \in \Gamma_0(t_1)\,\partial U$, contradicting (2.3). $\qquad\square$

The following proposition implies that B links A in Example 2.3.4 when $\dim M \geqslant 2$.

Proposition 2.5.6 *If A and B are disjoint closed bounded subsets of E such that $E \backslash A$ is path connected, then*

$$A \text{ links } B \implies B \text{ links } A.$$

Proof Suppose not, say, $\Gamma_0(B \times [0,1]) \cap A = \varnothing$ where $\Gamma_0 \in \Phi$. Let $u_0 = \Gamma_0(1) E$, and take $R > 0$ so large that $C = \{u \in E : \|u\| \leqslant R\} \supset A$. Since $E \backslash A$ is path connected, there is a path $\gamma \in C([0,1], E \backslash A)$ joining $u_0 = \gamma(0)$ to some point $\gamma(1) \in E \backslash C$. By *(iii)* in the definition of Φ,

$$\text{diam}(\Gamma_0(t_0) B - u_0) < \min\{\text{dist}(\gamma([0,1]), A), \text{dist}(\gamma(1), C)\}$$

for $t_0 < 1$ sufficiently close to 1. Then $\tilde{\gamma}(t) = \gamma((t - t_0)/(1 - t_0))$, $t \in [t_0, 1]$ satisfies $\tilde{\gamma}(t_0) = u_0$,

$$(\Gamma_0(t_0) B - u_0 + \tilde{\gamma}(t)) \cap A = \varnothing \quad \forall t \in [t_0, 1],$$

and

$$(\Gamma_0(t_0) B - u_0 + \tilde{\gamma}(1)) \cap C = \varnothing.$$

So

$$\Gamma_1(t) u = \begin{cases} \Gamma_0(t) u, & t \in [0, t_0] \\ \Gamma_0(t_0) u - u_0 + \tilde{\gamma}(t), & t \in (t_0, 1] \end{cases}$$

carries B outside C without intersecting A, and hence $B \cap \Gamma_1^{-1}(t) A = \varnothing$ for all $t \in [0,1]$ and $B \cap \Gamma_1^{-1}(1) C = \varnothing$. Let $\Gamma_2 \in \Phi$ be such that $\Gamma_2(C \times [0,1]) \subset C$. Then $\tilde{\Gamma} := \Gamma_1^{-1} \cdot \Gamma_2 \in \Phi$ and $B \cap \tilde{\Gamma}(A \times [0,1]) = \varnothing$, contradicting the fact that A links B. $\qquad\square$

Example 2.5.7 If $E = N \oplus M$, $u = v + w$ is a direct sum decomposition with $1 \leqslant \dim N < \infty$ and $\dim M \geqslant 2$, and $w_0 \in M$ with $\|w_0\| = 1$, then $B = \{w \in M : \|w\| = r\}$ links $A = \{v \in N : \|v\| \leqslant R\} \cup \{u = v + t w_0 : v \in N, t \geqslant 0, \|u\| = R\}$ for any $0 < r < R$.

Combining Theorem 2.5.3 and Proposition 2.5.6 now gives the following corollary.

Corollary 2.5.8 *Let A and B be closed bounded subsets of E such that A links B, $E \backslash A$ is path connected, and $\text{dist}(A, B) > 0$. If*

$$c := \inf_{\Gamma \in \Phi} \sup_{(u,t) \in A \times [0,1]} G(\Gamma(t) u), \qquad d := \sup_{\Gamma \in \Phi} \inf_{(u,t) \in B \times [0,1]} G(\Gamma(t) u)$$

are finite, $a := \sup G(A) \leqslant \inf G(B) =: b$, and G satisfies $(C)_c$ and $(C)_d$, then $c \geqslant b$ and $d \leqslant a$ are critical values of G.

2.6 Pairs of critical points with nontrivial critical groups

The following analog of Corollary 2.5.8 for homologically linking sets was obtained by Perera [104], where it was shown that the second critical point also has a nontrivial critical group. We assume that G has only a finite number of critical points throughout this section.

Theorem 2.6.1 *If A homologically links B in dimension $q \leqslant \dim E - 2$ and B is bounded, $G|_A \leqslant a < G|_B$ where a is a regular value, and G is bounded from below on bounded sets and satisfies (C), then G has two critical points u_1 and u_2 with*

$$G(u_1) > a > G(u_2), \quad C_{q+1}(G, u_1) \neq 0, \ C_q(G, u_2) \neq 0.$$

This is a special case of Theorem 2.6.3 below. When E is infinite dimensional the assumption that $q \leqslant \dim E - 2$ is, of course, satisfied. When $2 \leqslant d := \dim E < \infty$ it is related to the assumption in Corollary 2.5.8 that $E \backslash A$ is path connected. Indeed, suppose A homologically links B in dimension q and A is a compact neighborhood retract such that $E \backslash A$ is path connected. Since our coefficient group $\mathbb{Z}_2$ is a field, the vector space $\text{Hom}(H_{d-1}(A), \mathbb{Z}_2)$ of linear maps from $H_{d-1}(A)$ to $\mathbb{Z}_2$ is isomorphic to the singular cohomology group

$$\begin{aligned} H^{d-1}(A) &\approx \check{H}^{d-1}(A) && \text{since } A \text{ is a neighborhood retract} \\ &\approx H_1(E, E \backslash A) && \text{by the Poincaré duality theorem} \\ &\approx \tilde{H}_0(E \backslash A) && \text{as in Lemma 1.4.6 } (ii) \\ &= 0 && \text{since } E \backslash A \text{ is path connected,} \end{aligned}$$

where $\check{H}^*$ denotes Čech cohomology. So $H_{d-1}(A) = 0$ and hence $q \leqslant d - 2$. Combining Theorem 2.6.1 and Proposition 2.4.4 gives the following.

Corollary 2.6.2 *Let $E = N \oplus M$, $u = v + w$ be a direct sum decomposition with $d := \dim N < \infty$. If $G \leqslant a$ on $\{v \in N : \|v\| \leqslant R\} \cup \{u = v + tw_0 : v \in N, t \geqslant 0, \|u\| = R\}$ for some $R > 0$ and $w_0 \in M$ with $\|w_0\| = 1$ and $G > a$ on $\{w \in M : \|w\| = r\}$ for some $0 < r < R$, where a is a regular value, and G is bounded from below on bounded sets and satisfies (C), then G has two critical points u_1 and u_2 with*

$$G(u_1) > a > G(u_2), \quad C_{d+1}(G, u_1) \neq 0, \ C_d(G, u_2) \neq 0.$$

It was also shown by Perera [104] that the assumptions that B is bounded and G is bounded from below on bounded sets can be relaxed as follows; see also Schechter [139].

Theorem 2.6.3 *If A homologically links B in dimension q, $G|_A \leqslant a < G|_B$ where a is a regular value, and G is bounded from below on a set $C \supset B$ such that the inclusion-induced homomorphism $\tilde{H}_q(E \backslash C) \to \tilde{H}_q(E \backslash B)$ is trivial and satisfies (C), then G has two critical points u_1 and u_2 with*

$$G(u_1) > a > G(u_2), \quad C_{q+1}(G, u_1) \neq 0, \ C_q(G, u_2) \neq 0.$$

Proof Theorem 2.4.8 gives the critical point u_1. Let $c < \min \{a, \inf G(C)\}$ be a regular value. Then $G^c \subset G^a \subset E \backslash B$ and $G^c \subset E \backslash C \subset E \backslash B$, so we have the commutative diagram

$$
\begin{array}{ccc}
\tilde{H}_q(G^c) & \xrightarrow{\ i_* \ } \tilde{H}_q(G^a) \xrightarrow{\ j_* \ } H_q(G^a, G^c) \\
\downarrow & \quad \downarrow{\scriptstyle k_*} \\
\tilde{H}_q(E \backslash C) & \xrightarrow{\ l_* \ } \tilde{H}_q(E \backslash B)
\end{array}
$$

where the top row is a part of the exact sequence of the pair (G^a, G^c). We have $l_* = 0$, and $k_* \neq 0$ as in the proof of Proposition 2.4.7, so i_* is not onto. Then $j_* \neq 0$ by exactness and hence $H_q(G^a, G^c) \neq 0$, so G has a second critical point u_2 with $c < G(u_2) < a$ and $C_q(G, u_2) \neq 0$ by Proposition 1.4.2. $\square$

Proof of Theorem 2.6.1 Apply Theorem 2.6.3 with $C = \{u \in E : \|u\| \leqslant R\}$ for sufficiently large $R > 0$, noting that $\tilde{H}_q(E \backslash C) = 0$ since $q \leqslant \dim E - 2$. $\square$

Corollary 2.6.4 *Let $E = N \oplus M$, $u = v + w$ be a direct sum decomposition with $1 \leqslant d := \dim N < \infty$. If $G \leqslant a$ on $\{v \in N : \|v\| = R\}$ for some $R > 0$ and $G > a$ on M, where a is a regular value, and G is bounded from below on $\{tv_0 + w : t \geqslant 0, w \in M\}$ for some $v_0 \in N \backslash \{0\}$ and satisfies (C), then G has two critical points u_1 and u_2 with*

$$G(u_1) > a > G(u_2), \quad C_d(G, u_1) \neq 0, \ C_{d-1}(G, u_2) \neq 0.$$

Proof Apply Theorem 2.6.3 with A and B as in Proposition 2.4.3 and $C = \{tv_0 + w : t \geqslant 0, w \in M\}$, noting that the map

$$(E \backslash C) \times [0, 1] \to E \backslash C, \quad (u, t) \mapsto (1 - t) u - tv_0$$

is a contraction of $E \backslash C$ to $-v_0$ and hence $\tilde{H}_{d-1}(E \backslash C) = 0$. $\square$

2.7 Nonstandard geometries

Note that when N is infinite dimensional in Example 2.3.3 the set A is contractible and therefore does not link B homologically or homotopically. However, it does so according to Schechter and Tintarev's definition of linking if M is finite dimensional. This is a consequence of the following.

Proposition 2.7.1 *Let A and B be disjoint subsets of E with A bounded. If there is a sequence (B_j) of subsets of E such that A links each B_j, and $B_j = B'_j \cup B''_j$ where $B'_j \subset B$ and $\mathrm{dist}(B''_j, 0) \to \infty$, then A links B. In particular, A links $\bigcup_j B'_j$.*

Proof Let $\Gamma \in \Phi$. For each j, $\Gamma(A \times (0, 1]) \cap B_j \neq \varnothing$. But for sufficiently large j,

$$\sup_{(u,t)\in A\times[0,1]} \|\Gamma(t)\,u\| < \mathrm{dist}(B''_j, 0)$$

by Remark 2.5.1, so $\Gamma(A \times (0, 1]) \cap B'_j \neq \varnothing$. Hence $\Gamma(A \times (0, 1]) \cap B \neq \varnothing$. $\qquad\square$

Corollary 2.7.2 *If $E = N \oplus M$, $u = v + w$ is a direct sum decomposition with N nontrivial and M finite dimensional, then $A = \{v \in N : \|v\| = R\}$ links $B = M$ for any $R > 0$.*

Proof Apply Proposition 2.7.1 with $B'_j = \{w \in M : \|w\| \leqslant j\}$ and $B''_j = \{u = tv_0 + w : t \geqslant 0,\ w \in M,\ \|u\| = j\}$ for some $v_0 \in N$ with $\|v_0\| = 1$, noting that A links $B_j = B'_j \cup B''_j$ for $j > R$ by Example 2.5.7. $\qquad\square$

Applying Theorem 2.5.3 with $B = \overline{\bigcup_j B'_j}$ in the setting of Proposition 2.7.1 now gives the following.

Theorem 2.7.3 *Let A be a bounded subset of E and (B_j) a sequence of subsets of E such that A links each $B_j = B'_j \cup B''_j$ where*

$$\inf_j \mathrm{dist}(A, B'_j) > 0, \qquad \mathrm{dist}(B''_j, 0) \to \infty.$$

If

$$c := \inf_{\Gamma\in\Phi} \sup_{(u,t)\in A\times[0,1]} G(\Gamma(t)\,u)$$

is finite,

$$a := \sup_{u\in A} G(u) \leqslant \inf_j \inf_{u\in B'_j} G(u) =: b,$$

and G satisfies $(C)_c$, then $c \geqslant b$ is a critical value of G.

Corollary 2.7.4 *Let* $E = N \oplus M$, $u = v + w$ *be a direct sum decomposition with* N *nontrivial and* M *finite dimensional. If*

$$c := \inf_{\Gamma \in \Phi} \sup_{(u,t) \in A \times [0,1]} G(\Gamma(t) u)$$

is finite where $A = \{v \in N : \|v\| = R\}$ *for some* $R > 0$, $a := \sup G(A) \leqslant \inf G(M) =: b$, *and* G *satisfies* $(C)_c$, *then* $c \geqslant b$ *is a critical value of* G.

Proposition 2.7.1, Theorem 2.7.3, and Corollaries 2.7.2 and 2.7.4 are due to Schechter [142]; see also Ribarska *et al.* [131]. An analog of Theorem 2.7.3 for homologically linking sets does not seem to be known, but we have the following related result, which gives critical points with nontrivial critical groups under nonstandard geometrical assumptions that do not involve a finite-dimensional closed loop. We assume that G has only a finite number of critical points for the rest of this section.

Theorem 2.7.5 *Let* (A_j) *be a sequence of subsets of* E *and* B *a subset of* E *such that each* $A_j = A'_j \cup A''_j$ *homologically links* B *in dimension* q, *where* $\text{dist}(A''_j, B) \to \infty$. *If* $G|_{A'_j} \leqslant a < G|_B$ *where* a *is a regular value,*

$$b := \sup_j \sup_{u \in A''_j} G(u) < \infty, \qquad (2.4)$$

and G *satisfies* (PS), *then* G *has a critical point* u_1 *with*

$$G(u_1) > a, \quad C_{q+1}(G, u_1) \neq 0.$$

If, in addition, $q \leqslant \dim E - 2$, B *is bounded, and* G *is bounded from below on bounded sets, then there is a second critical point* u_2 *with*

$$G(u_2) < a, \quad C_q(G, u_2) \neq 0.$$

First we prove a deformation lemma.

Lemma 2.7.6 *If* $-\infty < a < b < +\infty$, C *is a set containing* K^b_a, $\delta > 0$, *and* G *satisfies* $(PS)_c$ *for all* $c \in [a, b]$, *then there are an* $\varepsilon_0 > 0$ *and, for each* $\varepsilon \in (0, \varepsilon_0)$, *a map* $\eta \in C(E \times [0, 1], E)$ *such that, writing* $\eta(u, t) = \eta(t) u$,

- *(i)* $\eta(0) = id_E$,
- *(ii)* $\eta(t)$ *is a homeomorphism of* E *for all* $t \in [0, 1]$,
- *(iii)* $\eta(t)$ *is the identity outside* $A = G^{b+\varepsilon}_{a-\varepsilon} \backslash N_{\delta/2}(C)$ *for all* $t \in [0, 1]$,
- *(iv)* $\|\eta(t) u - u\| \leqslant 1/\varepsilon \quad \forall (u, t) \in E \times [0, 1]$,
- *(v)* $G(\eta(\cdot) u)$ *is nonincreasing for each* $u \in E$,
- *(vi)* $\eta(1) (G^b \backslash N_{\delta+1/\varepsilon}(C)) \subset G^a$.

First, we have another lemma.

Lemma 2.7.7 *If* $-\infty < a < b < +\infty$, *N is an open neighborhood of* K_a^b, $k > 0$, *and G satisfies* $(PS)_c$ *for all* $c \in [a, b]$, *then there is an* $\varepsilon_0 > 0$ *such that*

$$\inf_{u \in G_{a-\varepsilon}^{b+\varepsilon} \backslash N} \| G'(u) \| \geqslant k\varepsilon \quad \forall \varepsilon \in (0, \varepsilon_0).$$

Proof If not, there are sequences $\varepsilon_j \searrow 0$ and $u_j \in G_{a-\varepsilon_j}^{b+\varepsilon_j} \backslash N$ such that

$$\| G'(u_j) \| < k\varepsilon_j.$$

Then (u_j) has a subsequence that is a $(PS)_c$ sequence for some $c \in [a, b]$, which in turn has a subsequence converging to some $u \in K^c \backslash N = \varnothing$, a contradiction. $\qquad\square$

Proof of Lemma 2.7.6 By Lemma 2.7.7, there is an $\varepsilon_0 > 0$ such that for each $\varepsilon \in (0, \varepsilon_0)$,

$$\| G'(u) \| \geqslant 4(b-a)\varepsilon \quad \forall u \in A. \tag{2.5}$$

Let V be a pseudo-gradient vector field for G, $g \in \mathrm{Lip}_{\mathrm{loc}}(E, [0, 1])$ satisfy $g = 0$ outside A and $g = 1$ on $B = G_a^b \backslash N_\delta(C)$, and $\eta(t)u$, $0 \leqslant t < T(u) \leqslant \infty$ the maximal solution of

$$\dot{\eta} = -2(b-a)g(\eta) \frac{V(\eta)}{\|V(\eta)\|^2}, \quad t > 0, \qquad \eta(0)u = u \in E. \tag{2.6}$$

Since

$$\| \eta(t)u - u \| \leqslant 2(b-a) \int_0^t \frac{g(\eta(\tau)u)}{\|V(\eta(\tau)u)\|} \, d\tau$$

$$\leqslant 4(b-a) \int_0^t \frac{g(\eta(\tau)u)}{\|G'(\eta(\tau)u)\|} \, d\tau \qquad \text{by (1.1)}$$

$$\leqslant t/\varepsilon \qquad \text{by (2.5)},$$

$\|\eta(\cdot)u\|$ is bounded if $T(u) < \infty$, so $T(u) = \infty$ and (i)–(iv) follow.

By (2.6) and (1.1),

$$\frac{d}{dt}(G(\eta(t)u)) = (G'(\eta), \dot{\eta}) = -2(b-a)g(\eta) \frac{(G'(\eta), V(\eta))}{\|V(\eta)\|^2}$$

$$\leqslant -(b-a)g(\eta) \leqslant 0 \tag{2.7}$$

and hence (v) holds. To see that (vi) holds, let $u \in G^b \backslash N_{\delta+1/\varepsilon}(C)$ and suppose that $\eta(1)u \notin G^a$. Then $\eta(t)u \in G_a^b$ for all $t \in [0, 1]$, and $\eta(t)u \notin N_\delta(C)$ by *(iv)*. Thus, $\eta(t) u \in B$ and hence $g(\eta(t) u) = 1$ for all $t \in [0, 1]$, so (2.7) gives

$$G(\eta(1) u) \leqslant G(u) - (b - a) \leqslant a,$$

a contradiction. $\qquad\square$

Proof of Theorem 2.7.5 If $a \geqslant b$, this follows from Theorems 2.4.8 and 2.6.1 with $A =$ any A_j, so we assume that $a < b$. Applying Lemma 2.7.6 with $C = K_a^b \cup B$ and $\delta > 0$, let $\varepsilon > 0$ be sufficiently small and η the corresponding map.

We claim that $\text{dist}(A_j'', C) \to \infty$. Since $\text{dist}(A_j'', B) \to \infty$ by assumption, it suffices to show that $\text{dist}(A_j'', K_a^b) \to \infty$ to prove the claim. By (PS), K_a^b is compact and hence bounded, so it is, in fact, enough to show that $\text{dist}(A_j'', 0) \to \infty$. But this follows since

$$\text{dist}(A_j'', 0) \geqslant \text{dist}(A_j'', B) - \text{dist}(B, 0)$$

by the triangle inequality.

Fix j so large that $\text{dist}(A_j'', C) > \delta + 1/\varepsilon$. Then $A_j'' \subset G^b \backslash N_{\delta+1/\varepsilon}(C)$ by (2.4), so $\eta(1) A_j'' \subset G^a$ by *(vi)*. Since $A_j' \subset G^a$, $\eta(1) A_j' \subset G^a$ by *(v)*, and $\eta(1) B = B$ by *(iii)*, so $G|_{\eta(1) A_j} \leqslant a < G|_{\eta(1) B}$. By *(ii)* and Proposition 2.4.5, $\eta(1) A_j$ homologically links $\eta(1) B$ in dimension q, so the conclusions follow from Theorems 2.4.8 and 2.6.1. $\qquad\square$

The following corollaries of Theorem 2.7.5 were obtained by Perera and Schechter [118]; see also Perera and Schechter [114] and Lancelotti [65].

Corollary 2.7.8 *Let $E = N \oplus M$, $u = v + w$ be a direct sum decomposition with $d := \dim M < \infty$. If $G > a$ on N, where a is a regular value, and G is bounded from above on M and satisfies* (PS), *then G has a critical point u with*

$$G(u) > a, \quad C_d(G, u) \neq 0.$$

Proof Apply Theorem 2.7.5 with $A_j' = \varnothing$ and $A_j'' = \{w \in M : \|w\| = j\}$, noting that each $A_j = A_j' \cup A_j''$ homologically links $B = N$ in dimension $q = d - 1$ and $b \leqslant \sup G(M) < \infty$. $\qquad\square$

Corollary 2.7.9 *Let $E = N \oplus M$, $u = v + w$ be a direct sum decomposition with $d := \dim M < \infty$. If $G < a$ on N, where a is a regular value, and G is bounded from below on M and satisfies* (PS), *then G has a critical point u with*

$$G(u) < a, \quad C_d(-G, u) \neq 0.$$

Proof Apply Corollary 2.7.8 to $-G$. $\qquad\square$

Corollary 2.7.10 *Let $E = N \oplus M$, $u = v + w$ be a direct sum decomposition with $d := \dim M < \infty$. If $G < a$ on $\{v \in N : \|v\| = R\}$ for some $R > 0$ and $G \geqslant a$ on M, where a is a regular value, and G is bounded from below on $\{tv_0 + w : t \geqslant 0, w \in M\}$ for some $v_0 \in N$ with $\|v_0\| = 1$, bounded from above on bounded sets, and satisfies* (PS), *then G has two critical points u_1 and u_2 with*

$$G(u_1) < a < G(u_2), \quad C_{d+1}(-G, u_1) \neq 0, \ C_d(-G, u_2) \neq 0.$$

Proof Apply Theorem 2.7.5 to $-G$ with $A'_j = \{w \in M : \|w\| \leqslant j\}$ and $A''_j = \{u = tv_0 + w : t \geqslant 0, w \in M, \|u\| = j\}$, noting that $A_j = A'_j \cup A''_j$ homologically links $B = \{v \in N : \|v\| = R\}$ in dimension $q = d$ for $j > R$. $\qquad\square$

We believe that under the hypotheses of Corollary 2.7.4, G has a critical point u with $C_d(-G, u) \neq 0$ where $d = \dim M$. More specifically, we conjecture that if $E = N \oplus M$, $u = v + w$ is a direct sum decomposition with $\dim N \geqslant 1$ and $d := \dim M < \infty$, $G \leqslant a$ on $\{v \in N : \|v\| = R\}$ for some $R > 0$ and $G > a$ on M, where a is a regular value, and G satisfies (PS), then G has a critical point u with

$$G(u) > a, \quad C_d(-G, u) \neq 0.$$

Note that when E is a Hilbert space and G is C^2 this implies

$$m(-G, u) \leqslant d \leqslant m^*(-G, u)$$

via Corollary 1.1.11. It was shown in Perera and Schechter [114] that G has critical points $\underline{u}$ and $\overline{u}$, not necessarily distinct, such that

$$G(\underline{u}) \leqslant G(\overline{u}), \quad m(-G, \underline{u}) \leqslant d \leqslant m^*(-G, \overline{u})$$

when, in addition, ∇G is a Fredholm nonlinear map of index zero near the critical points of G.

3

Applications to semilinear problems

3.1 Introduction

Now we give some applications of the previously discussed Morse theoretic and linking methods to the semilinear elliptic boundary value problem

$$\begin{cases} -\Delta u = f(x, u) & \text{in } \Omega \\ \quad\ \ u = 0 & \text{on } \partial\Omega \end{cases} \tag{3.1}$$

where Ω is a bounded domain in $\mathbb{R}^n$, $n \geq 1$ and f is a Carathéodory function on $\Omega \times \mathbb{R}$ satisfying the growth condition

$$|f(x, t)| \leq C \left(|t|^{r-1} + 1 \right) \quad \forall (x, t) \in \Omega \times \mathbb{R} \tag{3.2}$$

for some $r \in [2, 2^*)$ and a constant $C > 0$. Here

$$2^* = \begin{cases} 2n/(n-2), & n > 2 \\ \infty, & n \leq 2 \end{cases}$$

is the critical Sobolev exponent.

Weak solutions of (3.1) coincide with critical points of the C^1-functional

$$G(u) = \int_\Omega |\nabla u|^2 - 2F(x, u), \quad u \in E = H_0^1(\Omega)$$

where

$$F(x, t) = \int_0^t f(x, s)\, ds$$

is the primitive of f and $H_0^1(\Omega)$ is the usual Sobolev space with the norm

$$\|u\| = \left(\int_\Omega |\nabla u|^2 \right)^{1/2}.$$

47

When verifying the (PS) or the (C) condition for G it suffices to check the boundedness of the sequence by

Lemma 3.1.1 *Every bounded sequence* $(u_j) \subset E$ *such that* $G'(u_j) \to 0$ *has a convergent subsequence.*

Proof Fix $2^*/(2^* - r + 1) < q < 2^*$, so that, denoting by $q' = q/(q - 1)$ the Hölder conjugate of q, $(r - 1) q' < 2^*$. Since (u_j) is bounded, a renamed subsequence converges to some u weakly in E and strongly in $L^q(\Omega)$ by the compactness of the Sobolev embedding $E \hookrightarrow L^q(\Omega)$. We claim that $u_j \to u$ in E. Since E is a Hilbert space, it suffices to show that $(u_j - u, u_j) \to 0$. We have

$$(u_j - u, u_j) = \frac{1}{2} \left(G'(u_j), u_j - u \right) + \int_\Omega f(x, u_j) (u_j - u).$$

Since $G'(u_j) \to 0$ and $(u_j - u)$ is bounded, $(G'(u_j), u_j - u) \to 0$. By (3.2),

$$\left| \int_\Omega f(x, u_j) (u_j - u) \right| \leqslant C \int_\Omega \left(|u_j|^{r-1} + 1 \right) |u_j - u|$$

$$\leqslant C \left(\|u_j\|_{L^{(r-1)q'}(\Omega)}^{r-1} + |\Omega|^{1/q'} \right) \|u_j - u\|_{L^q(\Omega)}$$

where $|\Omega|$ denotes the volume of Ω, and (u_j) is bounded in $L^{(r-1)q'}(\Omega)$ by the Sobolev embedding $E \hookrightarrow L^{(r-1)q'}(\Omega)$. $\qquad\qquad\qquad\qquad\qquad\qquad\square$

3.2 Local nature of critical groups

Let u_0 be an isolated critical point of G. Clearly, the critical groups of G at u_0 depend only on the values of G near u_0. However, when $n \geqslant 2$, E is not embedded in $L^\infty(\Omega)$ and it would seem that $C_*(G, u_0)$ depend on the values of f everywhere. In this section we show that this is not the case.

Set

$$\tilde{G}(u) = G(u + u_0) - G(u_0)$$

$$= \int_\Omega |\nabla u|^2 - 2 \left(F(x, u + u_0) - F(x, u_0) - f(x, u_0) u \right) + \left(G'(u_0), u \right)$$

$$= \int_\Omega |\nabla u|^2 - 2\tilde{F}(x, u),$$

where $\tilde{F}$ is the primitive of

$$\tilde{f}(x, t) = f(x, t + u_0(x)) - f(x, u_0(x)).$$

Then $\tilde{f}$ also satisfies the growth condition (3.2) since $u_0 \in L^\infty(\Omega)$ by a standard regularity argument, 0 is an isolated critical point of $\tilde{G}$,

$$C_*(\tilde{G}, 0) = C_*(G, u_0),$$

and the values of $\tilde{f}(x, t)$ near $t = 0$ depend only on the values of $f(x, t)$ near $t = u_0(x)$. Thus, we may assume without loss of generality that $u_0 = 0$.

Lemma 3.2.1 *Let* $\delta > 0$, $\vartheta : \mathbb{R} \to [-\delta, \delta]$ *be a smooth nondecreasing function such that* $\vartheta(t) = -\delta$ *for* $t \leqslant -\delta$, $\vartheta(t) = t$ *for* $-\delta/2 \leqslant t \leqslant \delta/2$, *and* $\vartheta(t) = \delta$ *for* $t \geqslant \delta$, *and set*

$$G_1(u) = \int_\Omega |\nabla u|^2 - 2F(x, \vartheta(u)), \quad u \in E.$$

If 0 is an isolated critical point of G, then it is also an isolated critical point of G_1 *and*

$$C_q(G, 0) \approx C_q(G_1, 0) \quad \forall q.$$

Proof We apply Theorem 1.4.4 to the family

$$G_s(u) = \int_\Omega |\nabla u|^2 - 2F(x, (1 - s)u + s\,\vartheta(u)), \quad u \in E, \ s \in [0, 1]$$

in a small ball B_ε, noting that $G_0 = G$. Lemma 3.1.1 implies that each G_s satisfies (PS) over B_ε, and it is easy to see that the map $[0, 1] \to C^1(B_\varepsilon, \mathbb{R})$, $s \mapsto G_s$ is continuous, so it only remains to show that for sufficiently small $\varepsilon > 0$, B_ε contains no critical point of G_s other than 0 for all $s \in [0, 1]$.

Suppose $u_j \to 0$, $G'_{s_j}(u_j) = 0$, $s_j \in [0, 1]$, and $u_j \neq 0$. Then

$$\begin{cases} -\Delta u_j = f_j(x, u_j) & \text{in } \Omega \\ u_j = 0 & \text{on } \partial\Omega \end{cases}$$

where

$$f_j(x, t) = (1 - s_j + s_j\,\vartheta'(t))\, f(x, (1 - s_j)t + s_j\,\vartheta(t))$$

also satisfies the growth condition (3.2) for some generic positive constant C independent of j. Passing to a subsequence, we may assume that $u_j \to 0$ strongly in $L^2(\Omega)$ and a.e. in Ω, s_j converges to some $s \in [0, 1]$, and hence $f_j(x, u_j) \to f(x, 0) = 0$ a.e. Thus, if $u_j \to 0$ in $L^q(\Omega)$, then $f_j(x, u_j) \to 0$ in $L^{q/(r-1)}(\Omega)$. Since

$$\|u_j\|_{L^{nq/(n(r-1)-2q)}(\Omega)} \leqslant C \|u_j\|_{H^{2,q/(r-1)}(\Omega)} \leqslant C \|f_j(x, u_j)\|_{L^{q/(r-1)}(\Omega)}$$

by the Sobolev and Calderón–Zygmund inequalities, then $u_j \to 0$ also in $L^{nq/(n(r-1)-2q)}(\Omega)$. Starting with $q = 2$, iterating until $q > n(r - 1)/2$, and using the Sobolev embedding $H^{2,q/(r-1)}(\Omega) \hookrightarrow C(\overline{\Omega})$ now gives $u_j \to 0$ in $C(\overline{\Omega})$. So for sufficiently large j, $|u_j| \leqslant \delta/2$ and hence $G'(u_j) = 0$, contradicting our assumption that 0 is an isolated critical point of G. $\qquad\square$

The following theorem is now immediate from Lemma 3.2.1 and the remarks at the beginning of the section.

Theorem 3.2.2 *Let f_i, $i = 1, 2$ be Carathéodory functions on $\Omega \times \mathbb{R}$ satisfying*

$$|f_i(x, t)| \leqslant C\left(|t|^{r-1} + 1\right) \quad \forall(x, t) \in \Omega \times \mathbb{R}$$

for some $r \in [2, 2^*)$ and $C > 0$, $F_i(x, t) = \displaystyle\int_0^t f_i(x, s)\, ds$, and set

$$G_i(u) = \int_\Omega |\nabla u|^2 - 2F_i(x, u), \quad u \in E.$$

If u_0 is an isolated critical point of G_1 and

$$f_1(x, u_0(x) + t) = f_2(x, u_0(x) + t) \quad \forall x \in \Omega, \ |t| \leqslant \delta$$

for some $\delta > 0$, then u_0 is also an isolated critical point of G_2 and

$$C_q(G_1, u_0) \approx C_q(G_2, u_0) \quad \forall q.$$

Remark 3.2.3 The homotopy argument used in the proof of Lemma 3.2.1 was adapted from Degiovanni *et al.* [46].

3.3 Critical groups at zero

In this section we compute the critical groups of G at zero under various assumptions on the behavior of the nonlinearity $f(x, t)$ near $t = 0$.

First we consider the linear case $f(x, t) = \lambda t$, $\lambda \in \mathbb{R}$, i.e. the eigenvalue problem

$$\begin{cases} -\Delta u = \lambda u & \text{in } \Omega \\ \quad u = 0 & \text{on } \partial\Omega. \end{cases} \tag{3.3}$$

The spectrum $\sigma(-\Delta)$ of the negative Laplacian consists of isolated eigenvalues λ_l, $l \geqslant 1$ of finite multiplicities satisfying

$$0 < \lambda_1 < \lambda_2 < \cdots < \lambda_l < \cdots,$$

and

$$\|u\|^2 \geqslant \lambda_1 \|u\|_{L^2(\Omega)}^2 \quad \forall u \in E. \tag{3.4}$$

Let E_l be the eigenspace of λ_l,

$$N_l = \bigoplus_{j=1}^{l} E_j, \qquad M_l = N_l^\perp.$$

Then

$$E = N_l \oplus M_l, \quad u = v + w \tag{3.5}$$

is an orthogonal decomposition with respect to both the inner product in E and the $L^2(\Omega)$-inner product, and

$$\|v\|^2 \leqslant \lambda_l \|v\|_{L^2(\Omega)}^2 \qquad \forall v \in N_l, \tag{3.6}$$

$$\|w\|^2 \geqslant \lambda_{l+1} \|w\|_{L^2(\Omega)}^2 \qquad \forall w \in M_l. \tag{3.7}$$

When $\lambda \notin \sigma(-\Delta)$, the origin is the only critical point of

$$G(u) = \int_\Omega |\nabla u|^2 - \lambda u^2, \quad u \in E,$$

so the critical groups $C_q(G, 0)$ are defined. Let

$$d_l = \dim N_l.$$

We have the following theorem.

Theorem 3.3.1 *Let $\lambda \in \mathbb{R} \backslash \sigma(-\Delta)$.*

(i) If $\lambda < \lambda_1$, then

$$C_q(G, 0) \approx \delta_{q0} \, \mathbb{Z}_2.$$

(ii) If $\lambda_l < \lambda < \lambda_{l+1}$, then

$$C_q(G, 0) \approx \delta_{q d_l} \, \mathbb{Z}_2.$$

Proof (i) By (3.4),

$$G(u) \geqslant \left(1 - \frac{\lambda}{\lambda_1}\right) \|u\|^2 \quad \forall u \in E,$$

so 0 is the unique global minimizer of G and hence Proposition 1.5.2 applies.

(ii) Taking $U = E$ in (1.24) gives

$$C_q(G, 0) = H_q(G^0, G^0 \backslash \{0\}).$$

By (3.6),

$$G(v) \leqslant -\left(\frac{\lambda}{\lambda_l} - 1\right) \|v\|^2 \leqslant 0 \quad \forall v \in N_l,$$

so $N_l \subset G^0$. On the other hand, by (3.7),

$$G(w) \geq \left(1 - \frac{\lambda}{\lambda_{l+1}} \right) \|w\|^2 > 0 \quad \forall w \in M_l \setminus \{0\}, \tag{3.8}$$

so $G^0 \setminus \{0\} \subset E \setminus M_l$. Referring to the decomposition (3.5), let

$$\eta(u, t) = v + (1 - t) w, \quad u \in G^0, \, t \in [0, 1]$$

and note that

$$G(\eta(u, t)) = G(v) + (1 - t)^2 G(w) \leq G(v) + G(w) = G(u) \leq 0$$

by orthogonality and (3.8). So η is a strong deformation retraction of G^0 onto N_l and its restriction to $G^0 \setminus \{0\}$ is a strong deformation retraction onto $N_l \setminus \{0\}$. Thus,

$$C_q(G, 0) \approx H_q(N_l, N_l \setminus \{0\}) = \delta_{q d_l} \mathbb{Z}_2. \qquad \square$$

In the general case we have the next theorem.

Theorem 3.3.2 *Assume that* (3.2) *holds and* 0 *is an isolated critical point of* G.

(*i*) *If there is a* $\delta > 0$ *such that*

$$\frac{f(x, t)}{t} \leq \lambda_1 \quad \forall x \in \Omega, \, 0 < |t| \leq \delta, \tag{3.9}$$

then $C_q(G, 0) \approx \delta_{q0} \mathbb{Z}_2$.
(*ii*) *If there is a* $\delta > 0$ *such that*

$$\lambda_l \leq \frac{f(x, t)}{t} \leq \lambda_{l+1} \quad \forall x \in \Omega, \, 0 < |t| \leq \delta, \tag{3.10}$$

then $C_q(G, 0) \approx \delta_{q d_l} \mathbb{Z}_2$.

Proof (*i*) In view of Theorem 3.2.2, we may assume that (3.9) holds for all $(x, t) \in \Omega \times (\mathbb{R} \setminus \{0\})$, so $2F(x, t) \leq \lambda_1 t^2$ and hence

$$G(u) \geq \|u\|^2 - \lambda_1 \|u\|_{L^2(\Omega)}^2 \geq 0 \quad \forall u \in E$$

by (3.4). So 0 is a global minimizer of G, which is isolated by assumption, and hence Proposition 1.5.2 applies.

(*ii*) Referring to the decomposition (3.5), let

$$G_1(u) = \int_\Omega |\nabla w|^2 - |\nabla v|^2, \quad u \in E$$

and note that $C_q(G_1, 0) \approx \delta_{qd_l} \mathbb{Z}_2$ by an argument similar to that in the proof of Theorem 3.3.1 (*ii*). We apply Theorem 1.4.4 to the family

$$G_s(u) = (1 - s) G(u) + s G_1(u), \quad u \in E, \ s \in [0, 1]$$

in a small ball B_ε. Lemma 3.1.1 implies that each G_s satisfies (PS) over B_ε, and it is easy to see that the map $[0, 1] \to C^1(B_\varepsilon, \mathbb{R})$, $s \mapsto G_s$ is continuous. By assumption, 0 is an isolated critical point of $G_0 = G$, and we will show that it is the only critical point of G_s for $s > 0$.

Let $\overline{u} = -v + w$. Then

$$\frac{1}{2} \left(G'(u), \overline{u} \right) = \|w\|^2 - \|v\|^2 - \int_\Omega f(x, u) \overline{u}, \qquad \frac{1}{2} \left(G'_1(u), \overline{u} \right) = \|u\|^2$$

by orthogonality. In view of Theorem 3.2.2, we may assume that (3.10) holds for all $(x, t) \in \Omega \times (\mathbb{R} \setminus \{0\})$, so when $u(x) \neq 0$,

$$\begin{aligned}
f(x, u) \overline{u} &= \frac{f(x, u)}{u} u\overline{u} \\
&\leq \begin{cases} \lambda_{l+1} (w^2 - v^2), & u\overline{u} \geq 0 \\ -\lambda_l (v^2 - w^2), & u\overline{u} < 0 \end{cases} \\
&\leq \lambda_{l+1} w^2 - \lambda_l v^2.
\end{aligned}$$

When $u(x) = 0$, $f(x, u(x)) = 0$ and $v(x) = -w(x)$, so this inequality still holds. So

$$\frac{1}{2} \left(G'(u), \overline{u} \right) \geq \|w\|^2 - \|v\|^2 - \lambda_{l+1} \|w\|_{L^2(\Omega)}^2 + \lambda_l \|v\|_{L^2(\Omega)}^2 \geq 0$$

by (3.6) and (3.7). Thus,

$$\left(G'_s(u), \overline{u} \right) = (1 - s) \left(G'(u), \overline{u} \right) + s \left(G'_1(u), \overline{u} \right) \geq 2s \|u\|^2,$$

and hence $u = 0$ if $G'_s(u) = 0$ and $s > 0$. $\qquad\square$

Since $\lambda_1 > 0$, Theorem 3.3.2 (*i*) gives the following corollary.

Corollary 3.3.3 *If* (3.2) *holds, 0 is an isolated critical point of G, and*

$$t f(x, t) \leq 0 \quad \forall x \in \Omega, \ |t| \leq \delta$$

for some $\delta > 0$, then $C_q(G, 0) \approx \delta_{q0} \mathbb{Z}_2$.

An important special case of Theorem 3.3.2 is the asymptotically linear case

$$f(x, t) = \lambda t + o(t) \quad \text{as } t \to 0, \quad \text{uniformly a.e.} \tag{3.11}$$

for some $\lambda \in \mathbb{R}$. Then we say that (3.1) is resonant at zero if $\lambda \in \sigma(-\Delta)$, otherwise it is nonresonant. We have a corollary, as follows.

Corollary 3.3.4 *Assume (3.2) and (3.11) with* $\lambda \in \mathbb{R} \backslash \sigma(-\Delta)$.

(*i*) *If* $\lambda < \lambda_1$, *then*

$$C_q(G, 0) \approx \delta_{q0} \mathbb{Z}_2.$$

(*ii*) *If* $\lambda_l < \lambda < \lambda_{l+1}$, *then*

$$C_q(G, 0) \approx \delta_{qd_l} \mathbb{Z}_2.$$

Proof It only remains to show that 0 is an isolated critical point of G. If not, there is a sequence $(u_j) \subset E \backslash \{0\}$, $\rho_j := \|u_j\| \to 0$ such that $G'(u_j) = 0$. Then setting $\widetilde{u}_j := u_j / \rho_j$ we have $\|\widetilde{u}_j\| = 1$, so a renamed subsequence of $(\widetilde{u}_j)$ converges to some $\widetilde{u}$ weakly in E, strongly in $L^2(\Omega)$, and a.e. in Ω. By (3.11),

$$0 = \frac{(G'(u_j), v)}{2\rho_j} = \int_\Omega \nabla \widetilde{u}_j \cdot \nabla v - \lambda \widetilde{u}_j v + \mathrm{o}(1) \|v\|, \quad v \in E,$$

and passing to the limit gives

$$\int_\Omega \nabla \widetilde{u} \cdot \nabla v - \lambda \widetilde{u} v = 0 \quad \forall v \in E,$$

so $\widetilde{u}$ solves (3.3). Taking $v = \widetilde{u}_j$ and passing to the limit gives $\lambda \|\widetilde{u}\|^2_{L^2(\Omega)} = 1$, so $\widetilde{u} \neq 0$. So $\lambda \in \sigma(-\Delta)$, a contradiction. □

Corollary 3.3.5 *Assume that (3.2) and (3.11) hold and 0 is an isolated critical point of* G.

(*i*) *If* $\lambda = \lambda_1$ *and there is a* $\delta > 0$ *such that*

$$\frac{f(x, t)}{t} \leqslant \lambda_1 \quad \forall x \in \Omega, \, 0 < |t| \leqslant \delta,$$

then $C_q(G, 0) \approx \delta_{q0} \mathbb{Z}_2$.

(*ii*) *If* $\lambda = \lambda_l$ *and there is a* $\delta > 0$ *such that*

$$\frac{f(x, t)}{t} \geqslant \lambda_l \quad \forall x \in \Omega, \, 0 < |t| \leqslant \delta,$$

or if $\lambda = \lambda_{l+1}$ *and there is a* $\delta > 0$ *such that*

$$\frac{f(x, t)}{t} \leqslant \lambda_{l+1} \quad \forall x \in \Omega, \, 0 < |t| \leqslant \delta,$$

then $C_q(G, 0) \approx \delta_{qd_l} \mathbb{Z}_2$.

Remark 3.3.6 Theorem 3.3.2 (*ii*) was proved by Li *et al.* [74].

Finally we consider the case with a concave nonlinearity

$$f(x, t) = \mu |t|^{\sigma-2} t + \mathrm{o}(|t|^{\sigma-1}) \quad \text{as } t \to 0, \quad \text{uniformly a.e.} \tag{3.12}$$

for some $\mu \neq 0$ and $\sigma \in (1, 2)$. We have the following theorem.

Theorem 3.3.7 *Assume that* (3.2) *and* (3.12) *hold and* 0 *is an isolated critical point of G.*

(*i*) *If* $\mu < 0$, *then*

$$C_q(G, 0) \approx \delta_{q0} \, \mathbb{Z}_2.$$

(*ii*) *If* $\mu > 0$, *then*

$$C_q(G, 0) = 0 \quad \forall q.$$

Proof (*i*) This is immediate from Corollary 3.3.3.

(*ii*) This follows from Proposition 3.3.8 below since (3.12) and (3.2) imply

$$F(x, t) = \frac{\mu}{\sigma} |t|^{\sigma} + o(|t|^{\sigma}), \qquad H(x, t) = \left(\frac{2}{\sigma} - 1\right) \mu \, |t|^{\sigma} + o(|t|^{\sigma})$$

and $\sigma < 2$. □

Let

$$H(x, t) = 2F(x, t) - tf(x, t).$$

Proposition 3.3.8 *If* (3.2) *holds,*

$$F(x, t) \geqslant c \, |t|^{\sigma} \quad \forall x \in \Omega, \ |t| \leqslant \delta \qquad (3.13)$$

and

$$H(x, t) > 0 \quad \forall x \in \Omega, \ 0 < |t| \leqslant \delta \qquad (3.14)$$

for some $\sigma \in (1, 2)$ *and* $c, \delta > 0$, *and* 0 *is an isolated critical point of G, then*

$$C_q(G, 0) = 0 \quad \forall q.$$

Proof By (3.13) and (3.2),

$$F(x, t) \geqslant c \, |t|^{\sigma} - C \, |t|^{r} \quad \forall (x, t) \in \Omega \times \mathbb{R} \qquad (3.15)$$

for some constant $C > 0$. Set

$$\tilde{f}(x, t) = \begin{cases} -f(x, -\delta) \dfrac{t}{\delta}, & t < -\delta \\[2mm] f(x, t), & |t| \leqslant \delta \\[2mm] f(x, \delta) \dfrac{t}{\delta}, & t > \delta, \end{cases} \qquad \tilde{F}(x, t) = \int_0^t \tilde{f}(x, s) \, ds$$

and apply Theorem 3.2.2 to G and

$$\tilde{G}(u) = \int_\Omega |\nabla u|^2 - 2\tilde{F}(x, u)$$

to get $C_*(G, 0) \approx C_*(\widetilde{G}, 0)$. Since $\widetilde{F} = F$ on $\Omega \times [-\delta, \delta]$, (3.13) holds with $\widetilde{F}$ in place of F also. A simple calculation shows that

$$\widetilde{H}(x, t) = 2\widetilde{F}(x, t) - t\widetilde{f}(x, t) = \begin{cases} H(x, -\delta), & t < -\delta \\ H(x, t), & |t| \leqslant \delta \\ H(x, \delta), & t > \delta, \end{cases}$$

so (3.14) implies $\widetilde{H} > 0$ on $\Omega \times (\mathbb{R} \backslash \{0\})$. Thus, we may assume without loss of generality that $H > 0$ on $\Omega \times (\mathbb{R} \backslash \{0\})$.

We have

$$C_q(G, 0) = H_q(G^0 \cap B, G^0 \cap B \backslash \{0\}),$$

where

$$B = \{u \in E : \|u\| \leqslant 1\}$$

is the unit ball in E. We will show that $G^0 \cap B$ is contractible to 0 and $G^0 \cap B \backslash \{0\}$ is a strong deformation retract of $B \backslash \{0\} \simeq \partial B =: S$. Since E is infinite dimensional and hence S is contractible, the conclusion will then follow.

For $u \in S$ and $0 < t \leqslant 1$,

$$G(tu) = t^2 - \int_\Omega 2F(x, tu) \leqslant t^\sigma \left(t^{2-\sigma} - 2c \|u\|_{L^\sigma(\Omega)}^\sigma + C t^{r-\sigma} \right)$$

for some constant $C > 0$ by (3.15) and the Sobolev embedding $E \hookrightarrow L^r(\Omega)$. Since $c > 0$ and $\sigma < 2 \leqslant r$, $G(tu) < 0$ for all sufficiently small t (depending on u). Since

$$\begin{aligned} \frac{d}{dt} \left(G(tu) \right) &= 2 \left(t - \int_\Omega f(x, tu) u \right) \\ &= \frac{2}{t} \left(G(tu) + \int_\Omega H(x, tu) \right) \\ &> \frac{2}{t} G(tu), \end{aligned} \tag{3.16}$$

$G(tu) \geqslant 0 \implies \dfrac{d}{dt} \left(G(tu) \right) > 0$. Thus, there is a unique $0 < T(u) \leqslant 1$ such that $G(tu) < 0$ for $0 < t < T(u)$, $G(T(u)u) \leqslant 0$, and $G(tu) > 0$ for $T(u) < t \leqslant 1$.

We claim that the map $T : S \to (0, 1]$ is continuous. By (3.16) and the implicit function theorem, T is C^1 on $\{u \in S : T(u) < 1\}$, so it suffices to show that if $u_j \to u$ and $T(u) = 1$, then $T(u_j) \to 1$. But for any $t < 1$, $G(tu_j) \to G(tu) < 0$, so $T(u_j) > t$ for j sufficiently large.

Thus,

$$G^0 \cap B = \{tu : u \in S, \ 0 \leqslant t \leqslant T(u)\}$$

and is radially contractible to 0, and

$$(B \backslash \{0\}) \times [0, 1] \to B \backslash \{0\},$$

$$(u, t) \mapsto \begin{cases} (1 - t) u + t \, T(\pi(u)) \, \pi(u), & u \in B \backslash G^0 \\ u, & u \in G^0 \cap B \backslash \{0\}, \end{cases}$$

where π is the radial projection onto S, is a strong deformation retraction of $B \backslash \{0\}$ onto $G^0 \cap B \backslash \{0\}$. $\qquad \square$

Remark 3.3.9 Theorem 3.3.7 was proved by Perera [104, 105].

3.4 Asymptotically linear problems

In this section we consider the solvability and the existence of nontrivial solutions of the problem (3.1) in the asymptotically linear case

$$f(x, t) = \lambda t - p(x, t) \tag{3.17}$$

for some $\lambda \in \mathbb{R}$ and a Carathéodory function p on $\Omega \times \mathbb{R}$ satisfying

$$|p(x, t)| \leqslant C \left(|t|^{\tau - 1} + 1 \right) \quad \forall (x, t) \in \Omega \times \mathbb{R} \tag{3.18}$$

for some $\tau \in [1, 2)$ and a constant $C > 0$. We say that (3.1) is resonant at infinity if $\lambda \in \sigma(-\Delta)$, otherwise it is nonresonant. We have

$$G(u) = \int_\Omega |\nabla u|^2 - \lambda u^2 + 2P(x, u), \quad u \in E \tag{3.19}$$

where $P(x, t) = \displaystyle\int_0^t p(x, s) \, ds$ satisfies

$$|P(x, t)| \leqslant C \left(|t|^\tau + 1 \right) \quad \forall (x, t) \in \Omega \times \mathbb{R} \tag{3.20}$$

for some constant $C > 0$, and

$$\frac{1}{2} \left(G'(u), v \right) = \int_\Omega \nabla u \cdot \nabla v - \lambda u v + p(x, u) \, v, \quad u, v \in E. \tag{3.21}$$

By (3.18), (3.20), and the Sobolev embedding $E \hookrightarrow L^\tau(\Omega)$,

$$\left| \int_\Omega p(x, u) \, v \right| \leqslant C \left(\|u\|^{\tau - 1} + 1 \right) \|v\| \quad \forall u, v \in E \tag{3.22}$$

and

$$\left| \int_{\Omega} P(x, u) \right| \leqslant C \left(\|u\|^{\tau} + 1 \right) \quad \forall u \in E \qquad (3.23)$$

for some constant $C > 0$. We assume that G has only a finite number of critical points throughout this section.

The following lemma is useful for verifying the boundedness of (PS) sequences.

Lemma 3.4.1 *If* $G'(u_j) \to 0$ *and* $\rho_j := \|u_j\| \to \infty$, *then a subsequence of* $\tilde{u}_j := u_j/\rho_j$ *converges to a nontrivial solution of* (3.3), *in particular,* $\lambda \in \sigma(-\Delta)$.

Proof Since $\|\tilde{u}_j\| = 1$, a renamed subsequence converges to some $\tilde{u}$ weakly in E, strongly in $L^2(\Omega)$, and a.e. in Ω. By (3.21), (3.22), and the assumption that $\tau < 2$,

$$\frac{(G'(u_j), v)}{2\rho_j} = \int_{\Omega} \nabla \tilde{u}_j \cdot \nabla v - \lambda \tilde{u}_j v + o(1) \|v\|, \quad v \in E, \qquad (3.24)$$

and passing to the limit gives

$$\int_{\Omega} \nabla \tilde{u} \cdot \nabla v - \lambda \tilde{u} v = 0 \quad \forall v \in E, \qquad (3.25)$$

so $\tilde{u}$ solves (3.3). Taking $v = \tilde{u}$ in (3.25) gives $\|\tilde{u}\|^2 = \lambda \|\tilde{u}\|^2_{L^2(\Omega)}$, and taking $v = \tilde{u}_j$ in (3.24) and passing to the limit gives $\lambda \|\tilde{u}\|^2_{L^2(\Omega)} = 1$, so $\|\tilde{u}\| = 1$. Since $\tilde{u}_j \rightharpoonup \tilde{u}$ and $\|\tilde{u}_j\| \to \|\tilde{u}\|$, $\tilde{u}_j \to \tilde{u}$. $\qquad\square$

First we consider the nonresonant case.

Lemma 3.4.2 *If* $\lambda \notin \sigma(-\Delta)$, *then every sequence* $(u_j) \subset E$ *such that* $G'(u_j) \to 0$ *has a convergent subsequence, in particular,* G *satisfies* (PS).

Proof Since $\lambda \notin \sigma(-\Delta)$, (u_j) is bounded by Lemma 3.4.1, so the conclusion follows from Lemma 3.1.1. $\qquad\square$

Theorem 3.4.3 *Assume* (3.17) *and* (3.18).

(i) *If* $\lambda < \lambda_1$, *then* (3.1) *has a solution* u *with*

$$C_q(G, u) \approx \delta_{q0} \mathbb{Z}_2.$$

(ii) *If* $\lambda_l < \lambda < \lambda_{l+1}$, *then* (3.1) *has a solution* u *with*

$$C_{d_l}(G, u) \neq 0.$$

Proof G satisfies (PS) by Lemma 3.4.2.

(*i*) By (3.20),

$$2P(x,t) \geqslant -(\lambda_1 - \lambda)t^2 - C \quad \forall (x,t) \in \Omega \times \mathbb{R}$$

for some constant $C > 0$, and hence

$$G(u) \geqslant \int_\Omega |\nabla u|^2 - \lambda_1 u^2 - C \geqslant -C|\Omega| \quad \forall u \in E$$

by (3.19) and (3.4). So Propositions 1.5.1 and 1.5.2 apply.

(*ii*) By (3.20),

$$-(\lambda_{l+1} - \lambda)t^2 - C \leqslant 2P(x,t) \leqslant (\lambda - \lambda_l)t^2 + C \quad \forall (x,t) \in \Omega \times \mathbb{R}$$

for some constant $C > 0$, and hence

$$G(v) \leqslant \int_\Omega |\nabla v|^2 - \lambda_l v^2 + C \leqslant C|\Omega| \quad \forall v \in N_l$$

by (3.19) and (3.6), and

$$G(w) \geqslant \int_\Omega |\nabla w|^2 - \lambda_{l+1} w^2 - C \geqslant -C|\Omega| \quad \forall w \in M_l$$

by (3.19) and (3.7). So Corollary 2.7.8 applies. $\qquad\square$

In the resonant case we assume that either

(p_l^-) $\lambda = \lambda_l$ and $\alpha_\pm(x) := \liminf\limits_{t\to\pm\infty} \dfrac{tp(x,t)}{|t|^\tau}$ satisfy

$$\int_{y>0} \alpha_+(x)|y|^\tau + \int_{y<0} \alpha_-(x)|y|^\tau > 0 \quad \forall y \in E_l \setminus \{0\},$$

or

(p_l^+) $\lambda = \lambda_l$ and $\alpha^\pm(x) := \limsup\limits_{t\to\pm\infty} \dfrac{tp(x,t)}{|t|^\tau}$ satisfy

$$\int_{y>0} \alpha^+(x)|y|^\tau + \int_{y<0} \alpha^-(x)|y|^\tau < 0 \quad \forall y \in E_l \setminus \{0\}.$$

Note that $\alpha_\pm, \alpha^\pm \in L^\infty(\Omega)$ by (3.18) and

$$\alpha_\pm(x) \leqslant \liminf_{t\to\pm\infty} \frac{\tau P(x,t)}{|t|^\tau} \leqslant \limsup_{t\to\pm\infty} \frac{\tau P(x,t)}{|t|^\tau} \leqslant \alpha^\pm(x). \qquad (3.26)$$

Lemma 3.4.4 *If (p_l^-) or (p_l^+) holds, then every sequence $(u_j) \subset E$ such that $G'(u_j) \to 0$ has a convergent subsequence, in particular, G satisfies* (PS).

Proof It suffices to show that (u_j) is bounded by Lemma 3.1.1, so suppose $\rho_j := \|u_j\| \to \infty$. Then a renamed subsequence of $\tilde{u}_j := u_j/\rho_j$ converges to some $y \in E_l \setminus \{0\}$ strongly in E and a.e. in Ω by Lemma 3.4.1. Writing $u_j = y_j + z_j \in E_l \oplus E_l^\perp$, then $\tilde{z}_j := z_j/\rho_j \to 0$.

By (3.21),

$$\frac{1}{2}\left(G'(u_j), v\right) = \int_\Omega \nabla z_j \cdot \nabla v - \lambda_l z_j v + p(x, u_j)v, \quad v \in E,$$

and taking $v = y_j/\rho_j^\tau$ gives

$$\left| \int_\Omega \frac{y_j\, p(x, u_j)}{\rho_j^\tau} \right| \leqslant \frac{\|G'(u_j)\|}{2\rho_j^{\tau-1}} \to 0$$

since $\tau \geqslant 1$. By (3.22),

$$\left| \int_\Omega \frac{z_j\, p(x, u_j)}{\rho_j^\tau} \right| \leqslant C\left(1 + \frac{1}{\rho_j^{\tau-1}}\right) \|\tilde{z}_j\| \to 0.$$

Thus,

$$\int_\Omega \frac{u_j\, p(x, u_j)}{|u_j|^\tau}\, |\tilde{u}_j|^\tau = \int_\Omega \frac{u_j\, p(x, u_j)}{\rho_j^\tau} \to 0.$$

Since $u_j = \rho_j \tilde{u}_j \to \pm\infty$ a.e. on $\{y \gtrless 0\}$, then

$$\int_{y>0} \alpha_+(x)\,|y|^\tau + \int_{y<0} \alpha_-(x)\,|y|^\tau \leqslant 0 \leqslant \int_{y>0} \alpha^+(x)\,|y|^\tau + \int_{y<0} \alpha^-(x)\,|y|^\tau$$

by Fatou's lemma, contradicting (p_l^-) and (p_l^+). $\qquad\qquad\square$

Theorem 3.4.5 *Assume* (3.17) *and* (3.18).

(i) *If* (p_0^-) *holds, then* (3.1) *has a solution* u *with*

$$C_q(G, u) \approx \delta_{q0}\, \mathbb{Z}_2.$$

(ii) *If* (p_l^+) *or* (p_{l+1}^-) *holds, then* (3.1) *has a solution* u *with*

$$C_{d_l}(G, u) \neq 0.$$

Proof G satisfies (PS) by Lemma 3.4.4.

(i) We claim that every sequence $(u_j) \subset E$, $\rho_j := \|u_j\| \to \infty$ has a renamed subsequence such that $G(u_j) \to +\infty$, so that G is bounded from below and Propositions 1.5.1 and 1.5.2 apply. Writing $\tilde{u}_j := u_j/\rho_j = y_j/\rho_j + w_j/\rho_j =:$

$\widetilde{y}_j + \widetilde{w}_j \in E_1 \oplus M_1$, we have

$$G(u_j) = \int_\Omega |\nabla w_j|^2 - \lambda_1 \, w_j^2 + 2P(x, u_j)$$

$$\geqslant \rho_j^2 \left(1 - \frac{\lambda_1}{\lambda_2} \right) \|\widetilde{w}_j\|^2 + \int_\Omega 2P(x, u_j) \qquad (3.27)$$

by (3.19) and (3.7). Since

$$\left| \int_\Omega P(x, u_j) \right| \leqslant C \left(\rho_j^\tau + 1 \right)$$

by (3.23) and $\tau < 2$, if $\underline{\lim} \, \|\widetilde{w}_j\| > 0$, then $\underline{\lim} \, G(u_j)/\rho_j^2 > 0$ and the claim follows, so suppose $\widetilde{w}_j \to 0$ for a renamed subsequence. Since $\|\widetilde{y}_j\| \leqslant \|\widetilde{u}_j\| = 1$ and E_1 is finite dimensional, then a further subsequence of $(\widetilde{u}_j)$ converges to some $y \in E_1 \backslash \{0\}$ strongly in E and a.e. in Ω. Then $u_j = \rho_j \, \widetilde{u}_j \to \pm\infty$ a.e. on $\{y \gtrless 0\}$, so

$$\underline{\lim} \, \frac{G(u_j)}{\rho_j^\tau} \geqslant \underline{\lim} \int_\Omega \frac{2P(x, u_j)}{|u_j|^\tau} |\widetilde{u}_j|^\tau \qquad \text{by (3.27)}$$

$$\geqslant \frac{2}{\tau} \left(\int_{y>0} \alpha_+(x) |y|^\tau + \int_{y<0} \alpha_-(x) |y|^\tau \right) \qquad \begin{array}{l}\text{by Fatou's lemma} \\ \text{and (3.26)}\end{array}$$

$$> 0 \qquad \text{by } (p_0^-).$$

(*ii*) We will show that G is bounded from above on N_l and from below on M_l, so that Corollary 2.7.8 applies.

If (p_l^+) holds,

$$2P(x, t) \geqslant -(\lambda_{l+1} - \lambda_l) \, t^2 - C \quad \forall (x, t) \in \Omega \times \mathbb{R}$$

for some constant $C > 0$ by (3.20), and hence

$$G(w) \geqslant \int_\Omega |\nabla w|^2 - \lambda_{l+1} \, w^2 - C \geqslant -C \, |\Omega| \quad \forall w \in M_l$$

by (3.19) and (3.7). We claim that every sequence $(u_j) \subset N_l$, $\rho_j := \|u_j\| \to \infty$ has a renamed subsequence such that $G(u_j) \to -\infty$, so that G is bounded from above on N_l. Writing $\widetilde{u}_j := u_j/\rho_j = v_j/\rho_j + y_j/\rho_j =: \widetilde{v}_j + \widetilde{y}_j \in N_{l-1} \oplus E_l$, we have

$$G(u_j) = \int_\Omega |\nabla v_j|^2 - \lambda_l \, v_j^2 + 2P(x, u_j)$$

$$\leqslant -\rho_j^2 \left(\frac{\lambda_l}{\lambda_{l-1}} - 1 \right) \|\widetilde{v}_j\|^2 + \int_\Omega 2P(x, u_j) \qquad (3.28)$$

by (3.6). Since

$$\left| \int_{\Omega} P(x, u_j) \right| \leq C \left(\rho_j^{\tau} + 1 \right)$$

by (3.23) and $\tau < 2$, if $\underline{\lim} \, \|\tilde{v}_j\| > 0$, then $\overline{\lim} \, G(u_j)/\rho_j^2 < 0$ and the claim follows, so suppose $\tilde{v}_j \to 0$ for a renamed subsequence. Since $\|\tilde{y}_j\| \leq \|\tilde{u}_j\| = 1$ and E_l is finite dimensional, then a further subsequence of $(\tilde{u}_j)$ converges to some $y \in E_l \setminus \{0\}$ strongly in E and a.e. in Ω. Then $u_j = \rho_j \, \tilde{u}_j \to \pm\infty$ a.e. on $\{y \gtrless 0\}$, so

$$\overline{\lim} \, \frac{G(u_j)}{\rho_j^{\tau}} \leq \overline{\lim} \int_{\Omega} \frac{2P(x, u_j)}{|u_j|^{\tau}} \, |\tilde{u}_j|^{\tau} \qquad \text{by (3.28)}$$

$$\leq \frac{2}{\tau} \left(\int_{y>0} \alpha^+(x) \, |y|^{\tau} + \int_{y<0} \alpha^-(x) \, |y|^{\tau} \right) \qquad \begin{array}{l} \text{by Fatou's lemma} \\ \text{and (3.26)} \end{array}$$

$$< 0 \qquad \text{by } (p_l^+).$$

If (p_{l+1}^-) holds,

$$2P(x, t) \leq (\lambda_{l+1} - \lambda_l) \, t^2 + C \quad \forall (x, t) \in \Omega \times \mathbb{R}$$

for some constant $C > 0$ by (3.20), and hence

$$G(v) \leq \int_{\Omega} |\nabla v|^2 - \lambda_l \, v^2 + C \leq C \, |\Omega| \quad \forall v \in N_l$$

by (3.19) and (3.6). We claim that every sequence $(u_j) \subset M_l$, $\rho_j := \|u_j\| \to \infty$ has a renamed subsequence such that $G(u_j) \to +\infty$, so that G is bounded from below on M_l. Writing $\tilde{u}_j := u_j/\rho_j = y_j/\rho_j + w_j/\rho_j =: \tilde{y}_j + \tilde{w}_j \in E_{l+1} \oplus M_{l+1}$, we have

$$G(u_j) = \int_{\Omega} |\nabla w_j|^2 - \lambda_{l+1} \, w_j^2 + 2P(x, u_j)$$

$$\geq \rho_j^2 \left(1 - \frac{\lambda_{l+1}}{\lambda_{l+2}} \right) \|\tilde{w}_j\|^2 + \int_{\Omega} 2P(x, u_j) \qquad (3.29)$$

by (3.7). Since

$$\left| \int_{\Omega} P(x, u_j) \right| \leq C \left(\rho_j^{\tau} + 1 \right)$$

by (3.23) and $\tau < 2$, if $\underline{\lim} \, \|\tilde{w}_j\| > 0$, then $\underline{\lim} \, G(u_j)/\rho_j^2 > 0$ and the claim follows, so suppose $\tilde{w}_j \to 0$ for a renamed subsequence. Since $\|\tilde{y}_j\| \leq \|\tilde{u}_j\| = 1$ and E_{l+1} is finite dimensional, then a further subsequence of $(\tilde{u}_j)$ converges to some $y \in E_{l+1} \setminus \{0\}$ strongly in E and a.e. in Ω. Then $u_j = \rho_j \tilde{u}_j \to \pm\infty$ a.e. on $\{y \gtrless 0\}$, so

$$\varliminf \frac{G(u_j)}{\rho_j^\tau} \geq \varliminf \int_\Omega \frac{2P(x, u_j)}{|u_j|^\tau} |\tilde{u}_j|^\tau \qquad \text{by (3.29)}$$

$$\geq \frac{2}{\tau} \left(\int_{y>0} \alpha_+(x) |y|^\tau + \int_{y<0} \alpha_-(x) |y|^\tau \right) \qquad \begin{array}{l} \text{by Fatou's lemma} \\ \text{and (3.26)} \end{array}$$

$$> 0 \qquad \text{by } (p_{l+1}^-). \qquad \square$$

Set $d_0 = 0$. By Theorems 3.4.3 and 3.4.5, (3.1) has a solution u with $C_{d_l}(G, u) \neq 0$ under the condition

$$(f^l) \quad \begin{cases} \lambda < \lambda_1, \text{ or } (p_0^-) \text{ holds}, & l = 0 \\ \\ \lambda_l < \lambda < \lambda_{l+1}, \text{ or } (p_l^+) \text{ or } (p_{l+1}^-) \text{ holds}, & l \geq 1. \end{cases}$$

By Theorem 3.3.2, $C_q(G, 0) = 0$ for $q \neq d_{l_0}$ under the condition

(f_{l_0}) there is a $\delta > 0$ such that

$$\begin{cases} \dfrac{f(x, t)}{t} \leq \lambda_1 \quad \forall x \in \Omega, \ 0 < |t| \leq \delta, & l_0 = 0 \\ \\ \lambda_{l_0} \leq \dfrac{f(x, t)}{t} \leq \lambda_{l_0+1} \quad \forall x \in \Omega, \ 0 < |t| \leq \delta, & l_0 \geq 1. \end{cases}$$

Thus, we have the following theorem.

Theorem 3.4.6 *If (3.17), (3.18), (f^l), and (f_{l_0}) hold with $l_0 \neq l$, then (3.1) has a nontrivial solution.*

Remark 3.4.7 The nonresonant case of Theorem 3.4.6 is due to Amann and Zehnder [3]. Related results can be found in Bartsch and Li [14], Costa and Silva [33], Hirano and Nishimura [59], Lazer and Solimini [67], Li and Liu [71, 73], Li and Willem [76], Li and Zhang [78], Li and Zou [79], Perera [108], Schechter [141], Silva [150], Su and Tang [154], and Zou and Liu [162].

3.5 Problems with concave nonlinearities

In this section we consider the multiplicity of nontrivial solutions of (3.1) when f satisfies (3.17), (3.18), and

$$f(x, t) = \mu |t|^{\sigma-2} t + \lambda_0 t + o(t) \text{ as } t \to 0, \quad \text{uniformly a.e.} \qquad (3.30)$$

for some $\mu \neq 0$, $\sigma \in (1, 2)$, and $\lambda_0 \in \mathbb{R}$. Set

$$q(x, t) = f(x, t) - \mu |t|^{\sigma - 2} t, \qquad Q(x, t) = \int_0^t q(x, s) \, ds,$$

so that

$$G(u) = \int_\Omega |\nabla u|^2 - \frac{2\mu}{\sigma} |u|^\sigma - 2Q(x, u), \quad u \in E. \tag{3.31}$$

Again we assume that G has only a finite number of critical points. For simplicity we only consider the nonresonant case. Let

$$d_l = \dim N_l.$$

Theorem 3.5.1 *Assume* (3.17), (3.18), *and* (3.30) *with* $\lambda \notin \sigma(-\Delta)$.

(i) *If* $\lambda_0 > \lambda_l > \lambda$ *and*

$$2Q(x, t) \leqslant \lambda_{l+1} t^2 \quad \forall (x, t) \in \Omega \times \mathbb{R} \tag{3.32}$$

for some l, *then there is a* $\mu_* < 0$ *such that* (3.1) *has two nontrivial solutions* u_1 *and* u_2 *with*

$$G(u_1) \geqslant 0 > G(u_2), \quad C_{d_l}(G, u_1) \neq 0, \ C_{d_l - 1}(G, u_2) \neq 0 \tag{3.33}$$

for all $\mu \in (\mu_*, 0)$.

(ii) *If* $\lambda_0 < \lambda_{l+1} < \lambda$ *and*

$$2Q(x, t) \geqslant \lambda_l t^2 \quad \forall (x, t) \in \Omega \times \mathbb{R} \tag{3.34}$$

for some l, *then there is a* $\mu^* > 0$ *such that* (3.1) *has two nontrivial solutions* u_1 *and* u_2 *with*

$$G(u_1) > 0 \geqslant G(u_2), \quad C_{d_l + 1}(G, u_1) \neq 0, \ C_{d_l}(G, u_2) \neq 0 \tag{3.35}$$

for all $\mu \in (0, \mu^*)$.

Proof G satisfies (PS) by Lemma 3.4.2.

(i) We apply Corollary 2.6.4 to G with $N = N_l$, $M = M_l$, and $v_0 \in E_l \setminus \{0\}$, noting that then $d = d_l$ and $\{tv_0 + w : t \geqslant 0, \ w \in M\} \subset M_{l-1}$. For all $\mu < 0$,

$$G(w) \geqslant \int_\Omega |\nabla w|^2 - \lambda_{l+1} w^2 \geqslant 0 \quad \forall w \in M_l$$

by (3.31), (3.32), and (3.7). Since $\lambda_l > \lambda$,

$$2P(x, t) \geqslant -(\lambda_l - \lambda) t^2 - C \quad \forall (x, t) \in \Omega \times \mathbb{R}$$

for some constant $C > 0$ by (3.20), and hence

$$G(w) \geq \int_\Omega |\nabla w|^2 - \lambda_l\, w^2 - C \geq -C\, |\Omega| \quad \forall w \in M_{l-1}$$

by (3.19). Fixing $0 < \varepsilon < \lambda_0/\lambda_l - 1$,

$$2F(x,t) \geq \frac{2\mu}{\sigma}\, |t|^\sigma + (1+\varepsilon)\,\lambda_l\, t^2 - C\, |t|^r \quad \forall (x,t) \in \Omega \times \mathbb{R}$$

for some $r \in (2, 2^*)$ and a constant $C > 0$ by (3.30), (3.17), and (3.18), and hence

$$G(v) \leq -\varepsilon\, \|v\|^2 + C\big(|\mu|\, \|v\|^\sigma + \|v\|^r\big) \quad \forall v \in N_l$$

by (3.6) and the Sobolev embedding theorem, so there are $R > 0$, $a < 0$, and $\mu_* < 0$ such that $G \leq a$ on $\{v \in N_l : \|v\| = R\}$ for all $\mu \in (\mu_*, 0)$. Taking a larger if necessary, we may assume that G has no critical values in $[a, 0)$, so Corollary 2.6.4 gives two critical points u_1 and u_2 satisfying (3.33); $u_1 \neq 0$ since $C_q(G, 0) \approx \delta_{q0}\, \mathbb{Z}_2$ by Theorem 3.3.7 (i), and $u_2 \neq 0$ since $G(0) = 0$.

(ii) We apply Corollary 2.7.10 to $-G$ with $N = M_l$, $M = N_l$, and $v_0 \in E_{l+1}$ with $\|v_0\| = 1$, noting that then $d = d_l$ and $\{tv_0 + w : t \geq 0,\ w \in M\} \subset N_{l+1}$. For all $\mu > 0$,

$$-G(v) \geq \int_\Omega -|\nabla v|^2 + \lambda_l\, v^2 \geq 0 \quad \forall v \in N_l$$

by (3.31), (3.34), and (3.6). Since $\lambda > \lambda_{l+1}$,

$$2P(x,t) \leq (\lambda - \lambda_{l+1})\, t^2 + C \quad \forall (x,t) \in \Omega \times \mathbb{R}$$

for some constant $C > 0$ by (3.20), and hence

$$-G(v) \geq \int_\Omega -|\nabla v|^2 + \lambda_{l+1}\, v^2 - C \geq -C\, |\Omega| \quad \forall v \in N_{l+1}$$

by (3.19). Fixing $0 < \varepsilon < 1 - \lambda_0/\lambda_{l+1}$,

$$2F(x,t) \leq \frac{2\mu}{\sigma}\, |t|^\sigma + (1-\varepsilon)\,\lambda_{l+1}\, t^2 + C\, |t|^r \quad \forall (x,t) \in \Omega \times \mathbb{R}$$

for some $r \in (2, 2^*)$ and a constant $C > 0$ by (3.30), (3.17), and (3.18), and hence

$$-G(w) \leq -\varepsilon\, \|w\|^2 + C\big(\mu\, \|w\|^\sigma + \|w\|^r\big) \quad \forall w \in M_l$$

by (3.7) and the Sobolev embedding theorem, so there are $R > 0$, $a < 0$, and $\mu^* > 0$ such that $-G < a$ on $\{w \in M_l : \|w\| = R\}$ for all $\mu \in (0, \mu^*)$. Taking a larger if necessary, we may assume that $-G$ has no critical values in $[a, 0)$, so Corollary 2.7.10 gives two critical points u_1 and u_2 satisfying (3.35); $u_1 \neq 0$

since $G(0) = 0$, and $u_2 \neq 0$ since $C_q(G, 0) = 0$ for all q by Theorem 3.3.7 (*ii*).   □

Remark 3.5.2 Theorem 3.5.1 is a special case of a result due to Perera and Schechter [118]. Related results can be found in Ambrosetti *et al.* [6], Ambrosetti *et al.* [7], de Paiva and Massa [43], Li and Wang [75], Li *et al.* [77], Moroz [94], Perera [104, 105], and Wu and Yang [159].

4

Fučík spectrum

4.1 Introduction

Let us recall some terminology concerning mappings between Hilbert spaces.

Definition 4.1.1 Let H, H' be Hilbert spaces.

(i) $f : H \to H'$ is bounded if it maps bounded sets into bounded sets.

(ii) $f : H \to H'$ is positive homogeneous of degree $\alpha > 0$ if

$$f(su) = s^\alpha f(u) \quad \forall u \in H, \; s \geq 0.$$

Taking $u = 0$ and $s = 0$ gives $f(0) = 0$. When $\alpha = 1$ we will simply say that f is positive homogeneous.

(iii) $f : H \to H$ is monotone if

$$(f(u) - f(v), u - v) \geq 0 \quad \forall u, v \in H.$$

(iv) $f \in C(H, H)$ is a potential operator if there is a $F \in C^1(H, \mathbb{R})$, called a potential for f, such that

$$F'(u) = f(u) \quad \forall u \in H.$$

Replacing F with $F - F(0)$ gives a potential F with $F(0) = 0$.

Let H be a Hilbert space with the inner product $(\cdot, \cdot)$ and the associated norm $\|\cdot\|$. We assume that there are positive homogeneous monotone potential operators $p, n \in C(H, H)$ such that

$$p(u) + n(u) = u, \quad (p(u), n(u)) = 0 \quad \forall u \in H. \tag{4.1}$$

67

We use the suggestive notation

$$u^+ = p(u), \qquad u^- = -n(u),$$

so that (4.1) becomes

$$u = u^+ - u^-, \qquad (u^+, u^-) = 0. \tag{4.2}$$

This implies

$$\|u\|^2 = \|u^+\|^2 + \|u^-\|^2, \tag{4.3}$$

in particular,

$$\|u^\pm\| \leqslant \|u\|. \tag{4.4}$$

Let A be a self-adjoint operator on H with the spectrum $\sigma(A) \subset (0, \infty)$ and A^{-1} compact. Then $\sigma(A)$ consists of isolated eigenvalues $\lambda_l, l \geqslant 1$ of finite multiplicities satisfying

$$0 < \lambda_1 < \lambda_2 < \cdots < \lambda_l < \cdots .$$

Moreover,

$$D = D(A^{1/2})$$

is a Hilbert space with the inner product

$$(u, v)_D = (A^{1/2}u, A^{1/2}v) = (Au, v)$$

and the associated norm

$$\|u\|_D = \|A^{1/2}u\| = (Au, u)^{1/2}.$$

We have

$$\|u\|_D^2 = (Au, u) \geqslant \lambda_1 (u, u) = \lambda_1 \|u\|^2 \qquad \forall u \in H,$$

so $D \hookrightarrow H$, and the embedding is compact since A^{-1} is a compact operator.

Let E_l be the eigenspace of λ_l,

$$N_l = \bigoplus_{j=1}^{l} E_j, \qquad M_l = N_l^\perp \cap D.$$

Then

$$D = N_l \oplus M_l$$

is an orthogonal decomposition with respect to both $(\cdot, \cdot)$ and $(\cdot, \cdot)_D$. Moreover,

$$\|v\|_D^2 = (Av, v) \leqslant \lambda_l (v, v) = \lambda_l \|v\|^2 \qquad \forall v \in N_l, \tag{4.5}$$

$$\|w\|_D^2 = (Aw, w) \geqslant \lambda_{l+1} (w, w) = \lambda_{l+1} \|w\|^2 \qquad \forall w \in M_l. \tag{4.6}$$

We assume that

$$w \in M_1 \setminus \{0\} \implies w^\pm \neq 0. \tag{4.7}$$

Now let $f \in C(D, H)$ and consider the equation

$$Au = f(u), \quad u \in D. \tag{4.8}$$

We say that (4.8) has a jumping nonlinearity at zero (resp. infinity) if

$$f(u) = bu^+ - au^- + o(\|u\|_D) \quad \text{as} \quad \|u\|_D \to 0 \text{ (resp. } \infty) \tag{4.9}$$

for some $(a, b) \in \mathbb{R}^2$. The asymptotic equation associated with (4.8) when (4.9) holds is

$$Au = bu^+ - au^-, \quad u \in D. \tag{4.10}$$

The Fučík spectrum $\Sigma(A)$ of A is the set of points $(a, b) \in \mathbb{R}^2$ such that (4.10) has a nontrivial solution.

Note that if $a = b = \lambda$, then (4.10) reduces to the eigenvalue problem

$$Au = \lambda u, \quad u \in D,$$

which has a nontrivial solution if and only if λ is one of the eigenvalues λ_l, so the points (λ_l, λ_l) are in $\Sigma(A)$.

4.2 Examples

Here are some concrete examples of the equation (4.10).

Example 4.2.1 Perhaps the best-known example is the semilinear elliptic boundary value problem

$$\begin{cases} -\Delta u = bu^+ - au^- & \text{in } \Omega \\ u = 0 & \text{on } \partial\Omega \end{cases} \tag{4.11}$$

where Ω is a bounded domain in $\mathbb{R}^n$, $n \geq 1$ and $u^\pm = \max\{\pm u, 0\}$ are the positive and negative parts of u, respectively. Here $H = L^2(\Omega)$, $D = H_0^1(\Omega)$ is the usual Sobolev space, and A is the inverse of the solution operator $S : H \to D$, $f \mapsto u = (-\Delta)^{-1} f$ of the problem

$$\begin{cases} -\Delta u = f(x) & \text{in } \Omega \\ u = 0 & \text{on } \partial\Omega. \end{cases}$$

Since the embedding $D \hookrightarrow H$ is compact, $A^{-1} = S$ is compact on H.

The Fučík spectrum $\Sigma(-\Delta)$ of $-\Delta$ was introduced by Dancer [37, 38] and Fučík [53], who recognized its significance for the solvability of the problem

$$\begin{cases} -\Delta u = f(x, u) & \text{in } \Omega \\ \quad\; u = 0 & \text{on } \partial\Omega \end{cases} \tag{4.12}$$

when f is a Carathéodory function on $\Omega \times \mathbb{R}$ satisfying

$$|f(x, t)| \leqslant C\,(|t| + 1) \quad \forall (x, t) \in \Omega \times \mathbb{R} \tag{4.13}$$

for some constant $C > 0$ and

$$f(x, t) = bt^+ - at^- + \mathrm{o}(t) \quad \text{as } |t| \to \infty, \quad \text{uniformly a.e.}$$

In the ODE case $n = 1$, Fučík showed that $\Sigma(-d^2/dx^2)$ consists of a sequence of hyperbolic-like curves passing through the points (λ_l, λ_l), with one or two curves going through each point. In the PDE case $n \geqslant 2$ also, $\Sigma(-\Delta)$ consists, at least locally, of curves emanating from the points (λ_l, λ_l); see Gallouët and Kavian [54], Ruf [136], Lazer and McKenna [66], Lazer [68], Các [22], Magalhães [86], Cuesta and Gossez [35], de Figueiredo and Gossez [42], Schechter [140], and Margulies and Margulies [87]. In particular, it was shown in Schechter [140] that in the square $(\lambda_{l-1}, \lambda_{l+1})^2$, $\Sigma(-\Delta)$ contains two strictly decreasing curves, which may coincide, such that the points in the square that are either below the lower curve or above the upper curve are not in $\Sigma(-\Delta)$, while the points between them may or may not belong to $\Sigma(-\Delta)$ when they do not coincide.

Example 4.2.2 Let $n \geqslant 3$ and consider the weighted problem

$$\begin{cases} -\Delta u = V(x)\,(bu^+ - au^-) & \text{in } \Omega \\ \quad\; u = 0 & \text{on } \partial\Omega \end{cases}$$

with the weight $V \in L^n(\Omega)$ positive a.e. Here $H = L^2(\Omega)$, $D = H_0^1(\Omega)$, and A is the inverse of the solution operator $S : H \to D$, $f \mapsto u = (-\Delta)^{-1}(Vf)$ of

$$\begin{cases} -\Delta u = V(x)\,f(x) & \text{in } \Omega \\ \quad\; u = 0 & \text{on } \partial\Omega. \end{cases}$$

This includes singular weights such as $V(x) = |x|^{-q}$, $0 < q < 1$.

Example 4.2.3 Consider the problem

$$
\begin{cases}
-\Delta u + u = 0 & \text{in } \Omega \\
\dfrac{\partial u}{\partial v} = bu^{+} - au^{-} & \text{on } \partial\Omega
\end{cases}
$$

where the parameters a, b appear in the boundary condition, $\partial\Omega$ is now assumed to be C^{1}, and $\partial/\partial v$ is the exterior normal derivative on $\partial\Omega$. Here $H = L^{2}(\partial\Omega)$, $D = H^{1}(\Omega)$, and A is the inverse of the solution operator $S : H \to D,\ f \mapsto u$ of

$$
\begin{cases}
-\Delta u + u = 0 & \text{in } \Omega \\
\dfrac{\partial u}{\partial v} = f(x) & \text{on } \partial\Omega.
\end{cases}
$$

Since the trace embedding $D \hookrightarrow H$ is compact, $A^{-1} = S$ is compact on H.

Example 4.2.4 Consider the dynamic boundary value problem

$$
\begin{cases}
-(u^{\Delta}(t))^{\Delta} = b\,(u^{\sigma}(t))^{+} - a\,(u^{\sigma}(t))^{-}, & t \in (a, b) \cap \mathbb{T} \\
u(a) = u(b) = 0
\end{cases}
\tag{4.14}
$$

where $\mathbb{T}$, called a time scale, is a nonempty closed subset of $[a, b]$ such that $\min \mathbb{T} = a$, $\max \mathbb{T} = b$,

$$
\sigma(t) = \inf \{ s \in \mathbb{T} : s > t \}
$$

is the forward jump operator,

$$
u^{\Delta}(t) = \lim_{\substack{s \to t \\ s \neq \sigma(t)}} \frac{u(\sigma(t)) - u(s)}{\sigma(t) - s}
$$

is the Δ-derivative of u, and

$$
u^{\sigma}(t) = u(\sigma(t)).
$$

In particular, the top equation in (4.14) is an ordinary differential equation when $\mathbb{T}$ is continuous and a difference equation when $\mathbb{T}$ is discrete. Here $H = L_{\Delta}^{2}((a, b) \cap \mathbb{T})$, $D = H_{0,\Delta}^{1}((a, b) \cap \mathbb{T})$, and A is the inverse of the solution operator $S : H \to D,\ f \mapsto u$ of the problem

$$
\begin{cases}
-(u^{\Delta}(t))^{\Delta} = f(t), & t \in (a, b) \cap \mathbb{T} \\
u(a) = u(b) = 0.
\end{cases}
$$

Since the embedding $D \hookrightarrow H$ is compact, $A^{-1} = S$ is compact on H. We refer to Agarwal *et al.* [1] for the details.

Note that in all the above examples

$$(-u)^{\pm} = u^{\mp} \quad \forall u \in H. \tag{4.15}$$

Then u solves (4.10) if and only if

$$A(-u) = a(-u)^+ - b(-u)^-$$

and hence $(a, b) \in \Sigma(A)$ if and only if $(b, a) \in \Sigma(A)$, so $\Sigma(A)$ is symmetric about the line $a = b$. We do not have this symmetry in the general theory we are developing here since we do not assume (4.15). Here is an example without symmetry.

Example 4.2.5 Let $\Omega_{\pm}$ be disjoint subdomains of Ω such that $\overline{\Omega}_+ \cup \overline{\Omega}_- = \overline{\Omega}$ and consider the problem

$$\begin{cases} -\Delta u = b\,\chi_{\Omega_+}(x)\,u + a\,\chi_{\Omega_-}(x)\,u & \text{in } \Omega \\ \quad u = 0 & \text{on } \partial\Omega \end{cases}$$

where $\chi_{\Omega_{\pm}}$ are the characteristic functions of $\Omega_{\pm}$, respectively. Here

$$p(u) = \chi_{\Omega_+}(x)\,u, \qquad n(u) = \chi_{\Omega_-}(x)\,u$$

and H, D, and A are as in Example 4.2.1.

4.3 Preliminaries on operators

Let us prove some basic properties of positive homogeneous and potential operators. Let H, H' be Hilbert spaces.

Our first proposition gives a growth condition for a continuous positive homogeneous operator.

Proposition 4.3.1 *If $f \in C(H, H')$ is positive homogeneous, then there is a constant $C > 0$ such that*

$$\|f(u)\| \leqslant C\,\|u\| \quad \forall u \in H,$$

in particular, f is bounded.

Proof If not, there is a sequence $(u_j) \subset H \setminus \{0\}$ such that

$$\|f(u_j)\| > j^2\,\|u_j\| \quad \forall j. \tag{4.16}$$

Let $\hat{u}_j = u_j / (j\,\|u_j\|)$. Then

$$\|\hat{u}_j\| = \frac{1}{j} \to 0,$$

but

$$\|f(\hat{u}_j)\| = \frac{\|f(u_j)\|}{j \|u_j\|} > j \to \infty$$

by (4.16). This contradicts the continuity of f at zero. ☐

Our next proposition gives a formula for the potential of an operator.

Proposition 4.3.2 *Let $f \in C(H, H)$ be a potential operator and $F \in C^1(H, \mathbb{R})$ its potential with $F(0) = 0$. Then*

$$F(u) = \int_0^1 (f(su), u) \, ds, \quad u \in H,$$

in particular, if f is bounded, then so is F. If f is positive homogeneous, then

$$F(u) = \frac{1}{2} (f(u), u), \quad u \in H,$$

in particular, F is positive homogeneous of degree 2, bounded, and there is a constant $C > 0$ such that

$$|F(u)| \leq C \|u\|^2 \quad \forall u \in H.$$

Proof We have

$$F(u) = \int_0^1 \frac{d}{ds} (F(su)) \, ds = \int_0^1 (F'(su), u) \, ds = \int_0^1 (f(su), u) \, ds.$$

The last integral equals

$$\int_0^1 s (f(u), u) \, ds = \frac{1}{2} (f(u), u)$$

if f is positive homogeneous. ☐

Finally we have a proposition, as follows.

Proposition 4.3.3 *If $f \in C(H, H)$ is a positive homogeneous potential operator and $F \in C^1(H, \mathbb{R})$ is its potential, then there is a constant $C > 0$ such that*

$$|F(u) - F(v)| \leq C (\|u\| + \|v\|) \|u - v\| \quad \forall u, v \in H. \tag{4.17}$$

Proof We have

$$F(u) - F(v) = \int_0^1 \frac{d}{ds} (F(su + (1 - s)v)) \, ds$$

$$= \int_0^1 (f(su + (1 - s)v), u - v) \, ds,$$

so

$$|F(u) - F(v)| \leqslant \int_0^1 \|f(su + (1-s)v)\| \|u - v\| \, ds$$

$$\leqslant \left(\int_0^1 C \|su + (1-s)v\| \, ds \right) \|u - v\|,$$

where C is as in Proposition 4.3.1. Since

$$\|su + (1-s)v\| \leqslant s \|u\| + (1-s) \|v\| \leqslant \|u\| + \|v\|,$$

(4.17) follows.   □

4.4 Variational formulation

It is easy to see that A is a potential operator with the potential

$$\frac{1}{2} (Au, u) = \frac{1}{2} \|u\|_D^2.$$

By Proposition 4.3.2 and (4.2), the potentials of p, n are

$$\frac{1}{2} (p(u), u) = \frac{1}{2} (u^+, u^+ - u^-) = \frac{1}{2} \|u^+\|^2,$$

$$\frac{1}{2} (n(u), u) = \frac{1}{2} (-u^-, u^+ - u^-) = \frac{1}{2} \|u^-\|^2,$$

respectively. Let

$$I(u, a, b) = \|u\|_D^2 - a \|u^-\|^2 - b \|u^+\|^2, \quad u \in D.$$

Then $I(\cdot, a, b) \in C^1(D, \mathbb{R})$ with

$$I'(u) = 2(Au + au^- - bu^+),$$

so the critical points of I coincide with the solutions of (4.10). Thus, $(a, b) \in \Sigma(A)$ if and only if $I(\cdot, a, b)$ has a nontrivial critical point.

Lemma 4.4.1 *If $I'(u_j) \to 0$ and $\|u_j\|_D = 1$, then a subsequence of (u_j) converges to a nontrivial critical point of I, in particular, $(a, b) \in \Sigma(A)$.*

Proof Since

$$u_j = A^{-1}(bu_j^+ - au_j^- + I'(u_j)/2) = A^{-1}(u_j'),$$

where (u_j') is bounded and A^{-1} is compact, u_j converges to some $u \in D$ for a renamed subsequence. Then $I'(u) = 0$ by the continuity of I', and $u \neq 0$ since $\|u\|_D = 1$.   □

Proposition 4.4.2 *If $(a, b) \notin \Sigma(A)$, then every sequence $(u_j) \subset D$ such that $I'(u_j) \to 0$ converges to zero, in particular, I satisfies* (PS).

Proof If $u_j \nrightarrow 0$, then setting $\rho_j := \|u_j\|_D$ we have $\inf \rho_j > 0$ for a renamed subsequence. Let $\tilde{u}_j := u_j / \rho_j$. Then

$$I'(\tilde{u}_j) = \frac{I'(u_j)}{\rho_j} \to 0$$

and $\|\tilde{u}_j\|_D = 1$, so $(a, b) \in \Sigma(A)$ by Lemma 4.4.1, contrary to assumption. $\square$

Proposition 4.4.3 $\Sigma(A)$ *is closed.*

Proof Let $(a_j, b_j) \in \Sigma(A)$ converge to (a, b) and let $u_j \neq 0$ satisfy

$$A u_j = b_j u_j^+ - a_j u_j^- =: u_j'. \tag{4.18}$$

Replacing u_j with $u_j / \|u_j\|_D$ if necessary, we may assume that $\|u_j\|_D = 1$ and hence (u_j') is bounded. Since A^{-1} is compact, then $u_j = A^{-1}(u_j')$ converges to some $u \in D$ with $\|u\|_D = 1$ for a renamed subsequence, and passing to the limit in (4.18) shows that u solves (4.10), so $(a, b) \in \Sigma(A)$. $\square$

4.5 Some estimates

In this section we derive some estimates for I and I'.

Let

$$\underline{\lambda} = \min \{a, b\}, \qquad \bar{\lambda} = \max \{a, b\}.$$

Then

$$\|u\|_D^2 - \bar{\lambda} \|u\|^2 \leqslant I(u) \leqslant \|u\|_D^2 - \underline{\lambda} \|u\|^2 \quad \forall u \in D \tag{4.19}$$

by (4.3).

Lemma 4.5.1 *There is a constant $C > 0$ such that for all $(a_1, b_1), (a_2, b_2) \in \mathbb{R}^2$, $u_1, u_2 \in D$,*

$$|I(u_1, a_1, b_1) - I(u_2, a_2, b_2)| \leqslant (\|u_1\|_D + \|u_2\|_D) \|u_1 - u_2\|_D$$

$$+ C \bar{\lambda}_1 (\|u_1\| + \|u_2\|) \|u_1 - u_2\|$$

$$+ (|a_1 - a_2| + |b_1 - b_2|) \|u_2\|^2, \tag{4.20}$$

where $\bar{\lambda}_1 = \max \{a_1, b_1\}$.

Proof We have

$$I(u_1, a_1, b_1) - I(u_2, a_2, b_2)$$

$$= \|u_1\|_D^2 - \|u_2\|_D^2 - (a_1 \|u_1^-\|^2 - a_2 \|u_2^-\|^2) - (b_1 \|u_1^+\|^2 - b_2 \|u_2^+\|^2)$$

$$= (\|u_1\|_D + \|u_2\|_D)(\|u_1\|_D - \|u_2\|_D) - a_1 (\|u_1^-\|^2 - \|u_2^-\|^2)$$

$$- b_1 (\|u_1^+\|^2 - \|u_2^+\|^2) - (a_1 - a_2) \|u_2^-\|^2 - (b_1 - b_2) \|u_2^+\|^2. \qquad (4.21)$$

Since $\|u^\pm\|^2$ are the potentials of the positive homogeneous operators $2p, 2n$, respectively, there is a constant $C > 0$ such that

$$\left| \|u_1^\pm\|^2 - \|u_2^\pm\|^2 \right| \leqslant C \left(\|u_1\| + \|u_2\| \right) \|u_1 - u_2\|$$

by Proposition 4.3.3. So (4.20) follows from (4.21), the triangle inequality, and (4.4). $\qquad\qquad\square$

The following lemma is where we use the monotonicity of the operators p, n.

Lemma 4.5.2 *For all* $(a_1, b_1), (a_2, b_2) \in \mathbb{R}^2$, $u_1, u_2 \in D$,

$$\underline{\lambda}_1 \|u_1 - u_2\|^2 - (|a_1 - a_2| + |b_1 - b_2|) \|u_2\| \|u_1 - u_2\|$$

$$\leqslant \|u_1 - u_2\|_D^2 - \frac{1}{2} \left(I'(u_1, a_1, b_1) - I'(u_2, a_2, b_2), u_1 - u_2 \right)$$

$$\leqslant \overline{\lambda}_1 \|u_1 - u_2\|^2 + (|a_1 - a_2| + |b_1 - b_2|) \|u_2\| \|u_1 - u_2\|$$

where $\underline{\lambda}_1 = \min \{a_1, b_1\}$, $\overline{\lambda}_1 = \max \{a_1, b_1\}$.

Proof We have

$$\|u_1 - u_2\|_D^2 - \frac{1}{2} \left(I'(u_1, a_1, b_1) - I'(u_2, a_2, b_2), u_1 - u_2 \right)$$

$$= (b_1 u_1^+ - a_1 u_1^- - b_2 u_2^+ + a_2 u_2^-, u_1 - u_2)$$

$$= b_1 (u_1^+ - u_2^+, u_1 - u_2) - a_1 (u_1^- - u_2^-, u_1 - u_2)$$

$$+ ((b_1 - b_2) u_2^+ - (a_1 - a_2) u_2^-, u_1 - u_2).$$

Since p, n are monotone operators,

$$(u_1^- - u_2^-, u_1 - u_2) \leqslant 0 \leqslant (u_1^+ - u_2^+, u_1 - u_2),$$

and

$$(u_1^+ - u_2^+, u_1 - u_2) - (u_1^- - u_2^-, u_1 - u_2) = \|u_1 - u_2\|^2,$$

so

$$\underline{\lambda}_1 \left\| u_1 - u_2 \right\|^2 \leqslant b_1 \left(u_1^+ - u_2^+, u_1 - u_2 \right) - a_1 \left(u_1^- - u_2^-, u_1 - u_2 \right) \leqslant \bar{\lambda}_1 \left\| u_1 - u_2 \right\|^2.$$

By the Schwarz inequality and (4.4),

$$\left| \left((b_1 - b_2) u_2^+ - (a_1 - a_2) u_2^-, u_1 - u_2 \right) \right| \leqslant \left(|a_1 - a_2| + |b_1 - b_2| \right) \left\| u_2 \right\| \left\| u_1 - u_2 \right\|.$$

$\square$

In the special case $(a_1, b_1) = (a_2, b_2) = (a, b)$, Lemmas 4.5.1 and 4.5.2 reduce, respectively, to the following.

Lemma 4.5.3 *There is a constant $C > 0$ such that for all $u_1, u_2 \in D$,*

$$\left| I(u_1) - I(u_2) \right| \leqslant \left(\left\| u_1 \right\|_D + \left\| u_2 \right\|_D \right) \left\| u_1 - u_2 \right\|_D + C \bar{\lambda} \left(\left\| u_1 \right\| + \left\| u_2 \right\| \right) \left\| u_1 - u_2 \right\|.$$

Lemma 4.5.4 *For all $u_1, u_2 \in D$,*

$$\underline{\lambda} \left\| u_1 - u_2 \right\|^2 \leqslant \left\| u_1 - u_2 \right\|_D^2 - \frac{1}{2} \left(I'(u_1) - I'(u_2), u_1 - u_2 \right) \leqslant \bar{\lambda} \left\| u_1 - u_2 \right\|^2.$$

4.6 Convexity and concavity

In this section we show that $I(v + y + w, a, b)$, $v + y + w \in N_{l-1} \oplus E_l \oplus M_l$ is strictly concave in v and strictly convex in w when (a, b) is in the square

$$Q_l = (\lambda_{l-1}, \lambda_{l+1})^2, \quad l \geqslant 2.$$

Proposition 4.6.1 *Let $(a, b) \in Q_l$.*

(i) *If $v_1 \neq v_2 \in N_{l-1}$, $w \in M_{l-1}$, then*

$$I((1 - t)v_1 + t v_2 + w) > (1 - t) I(v_1 + w) + t I(v_2 + w) \quad \forall t \in (0, 1).$$

(ii) *If $v \in N_l$, $w_1 \neq w_2 \in M_l$, then*

$$I(v + (1 - t)w_1 + t w_2) < (1 - t) I(v + w_1) + t I(v + w_2) \quad \forall t \in (0, 1).$$

Proof (i) We have

$$I((1 - t) v_1 + t v_2 + w) - (1 - t) I(v_1 + w) - t I(v_2 + w)$$

$$= (1 - t) \left[I((1 - t) v_1 + t v_2 + w) - I(v_1 + w) \right]$$

$$- t \left[I(v_2 + w) - I((1 - t) v_1 + t v_2 + w) \right]$$

$$= (1 - t) t \left(I'(u_1) - I'(u_2), v_2 - v_1 \right),$$

where $u_1 = (1 - t_1)\, v_1 + t_1\, v_2 + w$, $u_2 = (1 - t_2)\, v_1 + t_2\, v_2 + w$ for some $0 < t_1 < t < t_2 < 1$ by the mean-value theorem. Applying the first inequality in Lemma 4.5.4 gives

$$\frac{1}{2}\left(I'(u_1) - I'(u_2), v_2 - v_1\right)$$
$$\geq (t_2 - t_1)(\underline{\lambda}\, \|v_1 - v_2\|^2 - \|v_1 - v_2\|_D^2)$$
$$\geq (t_2 - t_1)(\underline{\lambda} - \lambda_{l-1})\, \|v_1 - v_2\|^2 \qquad \text{by (4.5)}$$
$$> 0$$

since $t_2 > t_1, \underline{\lambda} > \lambda_{l-1}$, and $v_1 \neq v_2$.

(*ii*) We have

$$I(v + (1 - t)\, w_1 + t\, w_2) - (1 - t)\, I(v + w_1) - t\, I(v + w_2)$$
$$= (1 - t)\left[I(v + (1 - t)\, w_1 + t\, w_2) - I(v + w_1)\right]$$
$$\quad - t\left[I(v + w_2) - I(v + (1 - t)\, w_1 + t\, w_2)\right]$$
$$= (1 - t)\, t\left(I'(u_1) - I'(u_2), w_2 - w_1\right),$$

where $u_1 = v + (1 - t_1)\, w_1 + t_1\, w_2$, $u_2 = v + (1 - t_2)\, w_1 + t_2\, w_2$ for some $0 < t_1 < t < t_2 < 1$ by the mean-value theorem. Applying the second inequality in Lemma 4.5.4 gives

$$\frac{1}{2}\left(I'(u_1) - I'(u_2), w_2 - w_1\right)$$
$$\leq -(t_2 - t_1)(\|w_1 - w_2\|_D^2 - \overline{\lambda}\, \|w_1 - w_2\|^2)$$
$$\leq -(t_2 - t_1)(\lambda_{l+1} - \overline{\lambda})\, \|w_1 - w_2\|^2 \qquad \text{by (4.6)}$$
$$< 0$$

since $t_2 > t_1, \lambda_{l+1} > \overline{\lambda}$, and $w_1 \neq w_2$. □

4.7 Minimal and maximal curves

In this section we construct the minimal and maximal curves of $\Sigma(A)$ in Q_l. The development here closely follows Schechter [140].

Proposition 4.7.1 *Let $(a, b) \in Q_l$.*

(*i*) *There is a positive homogeneous map $\theta(\cdot, a, b) \in C(M_{l-1}, N_{l-1})$ such that $v = \theta(w)$ is the unique solution of*

$$I(v + w) = \sup_{v' \in N_{l-1}} I(v' + w), \quad w \in M_{l-1}. \qquad (4.22)$$

Moreover,

$$I'(v + w) \perp N_{l-1} \iff v = \theta(w). \tag{4.23}$$

(ii) *There is a positive homogeneous map* $\tau(\cdot, a, b) \in C(N_l, M_l)$ *such that* $w = \tau(v)$ *is the unique solution of*

$$I(v + w) = \inf_{w' \in M_l} I(v + w'), \quad v \in N_l. \tag{4.24}$$

Moreover,

$$I'(v + w) \perp M_l \iff w = \tau(v). \tag{4.25}$$

Proof (i) For $v \in N_{l-1}$, $w \in M_{l-1}$,

$$
\begin{aligned}
I(v + w) &\leqslant \|v + w\|_D^2 - \underline{\lambda}\|v + w\|^2 && \text{by (4.19)} \\
&= \|v\|_D^2 + \|w\|_D^2 - \underline{\lambda}(\|v\|^2 + \|w\|^2) \\
&\leqslant -(\underline{\lambda} - \lambda_{l-1})\|v\|^2 + \|w\|_D^2 - \underline{\lambda}\|w\|^2 && \text{by (4.5).}
\end{aligned}
$$

Since $\underline{\lambda} > \lambda_{l-1}$, this implies that $I(\cdot + w)$ is bounded from above and anticoercive on the finite-dimensional space N_{l-1}, so (4.22) has a solution $v = \theta(w)$ satisfying

$$I'(\theta(w) + w) \perp N_{l-1}.$$

It is unique by Proposition 4.6.1 (i).

If $I'(v + w) \perp N_{l-1}$, then applying the first inequality in Lemma 4.5.4 with $u_1 = v + w$, $u_2 = \theta(w) + w$ and noting that $u_1 - u_2 = v - \theta(w) \perp I'(u_1) - I'(u_2)$ gives

$$\underline{\lambda}\|v - \theta(w)\|^2 \leqslant \|v - \theta(w)\|_D^2 \leqslant \lambda_{l-1}\|v - \theta(w)\|^2,$$

so $v = \theta(w)$.

Next we show that θ is bounded. Applying the first inequality in Lemma 4.5.4 with $u_1 = \theta(w) + w$, $u_2 = w$ and noting that $u_1 - u_2 = \theta(w) \perp I'(u_1)$ gives

$$
\begin{aligned}
\underline{\lambda}\|\theta(w)\|^2 &\leqslant \|\theta(w)\|_D^2 + \frac{1}{2}\left(I'(w), \theta(w)\right) \\
&\leqslant \lambda_{l-1}\|\theta(w)\|^2 + (aw^- - bw^+, \theta(w))
\end{aligned}
$$

since

$$(Aw, \theta(w)) = (w, \theta(w))_D = 0.$$

Since

$$\left|(aw^- - bw^+, \theta(w))\right| \leq (a + b) \|w\| \, \|\theta(w)\|,$$

it follows that

$$\|\theta(w)\| \leq \frac{2\overline{\lambda}}{\underline{\lambda} - \lambda_{l-1}} \|w\|. \tag{4.26}$$

Now to see that θ is continuous, let $w_0, w \in M_{l-1}$. Applying the first inequality in Lemma 4.5.4 with $u_1 = \theta(w) + w$, $u_2 = \theta(w_0) + w_0$ and noting that $\theta(w) - \theta(w_0) \perp I'(u_1) - I'(u_2)$ gives

$$\underline{\lambda} \left(\|\theta(w) - \theta(w_0)\|^2 + \|w - w_0\|^2 \right)$$

$$\leq \|\theta(w) - \theta(w_0)\|_D^2 + \|w - w_0\|_D^2$$

$$- \frac{1}{2} \left(I'(\theta(w) + w) - I'(\theta(w_0) + w_0), w - w_0 \right)$$

$$\leq \lambda_{l-1} \|\theta(w) - \theta(w_0)\|^2 - \left(a \, (\theta(w) + w)^- - b \, (\theta(w) + w)^+ \right.$$

$$- a \, (\theta(w_0) + w_0)^- + b \, (\theta(w_0) + w_0)^+, w - w_0 \big) \tag{4.27}$$

since

$$(A(\theta(w) - \theta(w_0)), w - w_0) = (\theta(w) - \theta(w_0), w - w_0)_D = 0$$

and

$$(A(w - w_0), w - w_0) = \|w - w_0\|_D^2.$$

Together with (4.26), (4.27) implies

$$\|\theta(w) - \theta(w_0)\|^2 \leq C \left(\|w\| + \|w_0\| \right) \|w - w_0\| \tag{4.28}$$

for some constant $C > 0$, so $\theta(w) \to \theta(w_0)$ as $w \to w_0$.

For $s \geq 0$, by (4.23),

$$I'(s \, \theta(w) + sw) = s \, I'(\theta(w) + w) \perp N_{l-1}$$

and hence

$$\theta(sw) = s \, \theta(w).$$

(*ii*) For $v \in N_l$, $w \in M_l$,

$$I(v + w) \geq \|v + w\|_D^2 - \overline{\lambda} \|v + w\|^2 \qquad \text{by (4.19)}$$

$$= \|v\|_D^2 + \|w\|_D^2 - \overline{\lambda} \left(\|v\|^2 + \|w\|^2 \right)$$

$$\geq (1 - \overline{\lambda}/\lambda_{l+1}) \|w\|_D^2 + \|v\|_D^2 - \overline{\lambda} \|v\|^2 \qquad \text{by (4.6).}$$

Since $\bar\lambda < \lambda_{l+1}$, this implies that $I(v + \cdot)$ is bounded from below and coercive on M_l. It is also weakly lower semicontinuous since the embedding $D \hookrightarrow H$ is compact. So (4.24) has a solution $w = \tau(v)$ satisfying

$$I'(v + \tau(v)) \perp M_l.$$

It is unique by Proposition 4.6.1 (*ii*).

If $I'(v + w) \perp M_l$, then applying the second inequality in Lemma 4.5.4 with $u_1 = v + w$, $u_2 = v + \tau(v)$ and noting that $u_1 - u_2 = w - \tau(v) \perp I'(u_1) - I'(u_2)$ gives

$$\lambda_{l+1} \|w - \tau(v)\|^2 \le \|w - \tau(v)\|_D^2 \le \bar\lambda \|w - \tau(v)\|^2 ,$$

so $w = \tau(v)$.

Next we show that τ is bounded. Applying the second inequality in Lemma 4.5.4 with $u_1 = v + \tau(v)$, $u_2 = v$ and noting that $u_1 - u_2 = \tau(v) \perp I'(u_1)$ gives

$$\|\tau(v)\|_D^2 \le \bar\lambda \|\tau(v)\|^2 - \frac{1}{2}\left(I'(v), \tau(v)\right)$$

$$\le \bar\lambda \|\tau(v)\|_D^2 /\lambda_{l+1} - (av^- - bv^+, \tau(v))$$

since

$$(Av, \tau(v)) = (v, \tau(v))_D = 0.$$

Since

$$\left|(av^- - bv^+, \tau(v))\right| \le (a + b) \|v\| \|\tau(v)\|_D /\lambda_1,$$

it follows that

$$\|\tau(v)\|_D \le \frac{2\bar\lambda}{\lambda_1 \left(1 - \bar\lambda/\lambda_{l+1}\right)} \|v\| . \tag{4.29}$$

Now to see that τ is continuous, let $v_0, v \in N_l$. Applying the second inequality in Lemma 4.5.4 with $u_1 = \tau(v) + v$, $u_2 = \tau(v_0) + v_0$ and noting that $\tau(v) - \tau(v_0) \perp I'(u_1) - I'(u_2)$ gives

$$\|\tau(v) - \tau(v_0)\|_D^2$$

$$\le \bar\lambda \left(\|v - v_0\|^2 + \|\tau(v) - \tau(v_0)\|^2\right) - \|v - v_0\|_D^2$$

$$+ \frac{1}{2}\left(I'(\tau(v) + v) - I'(\tau(v_0) + v_0), v - v_0\right)$$

$$\le \bar\lambda \|v - v_0\|^2 + \bar\lambda \|\tau(v) - \tau(v_0)\|_D^2 /\lambda_{l+1} + (a\left(\tau(v) + v\right)^-$$

$$- b\left(\tau(v) + v\right)^+ - a\left(\tau(v_0) + v_0\right)^- + b\left(\tau(v_0) + v_0\right)^+, v - v_0) \tag{4.30}$$

since
$$(A(\tau(v) - \tau(v_0)), v - v_0) = (\tau(v) - \tau(v_0), v - v_0)_D = 0$$

and
$$(A(v - v_0), v - v_0) = \|v - v_0\|_D^2 .$$

Together with (4.29), (4.30) implies
$$\|\tau(v) - \tau(v_0)\|_D^2 \leqslant C (\|v\| + \|v_0\|) \|v - v_0\|$$

for some constant $C > 0$, so $\tau(v) \to \tau(v_0)$ as $v \to v_0$.

For $s \geqslant 0$, by (4.25),
$$I'(sv + s\tau(v)) = s I'(v + \tau(v)) \perp M_l$$

and hence
$$\tau(sv) = s\tau(v). \qquad \qquad \Box$$

Next we show that θ, τ are locally bounded in and depend continuously on (a, b).

Lemma 4.7.2 *Given* $(a_0, b_0) \in Q_l$ *and a neighborhood* $N \subset\subset Q_l$ *of* (a_0, b_0), *there is a constant* $C > 0$ *such that*

(i) *for all* $w \in M_{l-1}$ *and* $(a, b) \in N$,
$$\|\theta(w, a, b)\| \leqslant C \|w\|_D , \qquad \qquad (4.31)$$
$$\|\theta(w, a, b) - \theta(w, a_0, b_0)\| \leqslant C (|a - a_0| + |b - b_0|) \|w\|_D , \quad (4.32)$$

(ii) *for all* $v \in N_l$ *and* $(a, b) \in N$,
$$\|\tau(v, a, b)\|_D \leqslant C \|v\| , \qquad \qquad (4.33)$$
$$\|\tau(v, a, b) - \tau(v, a_0, b_0)\|_D \leqslant C (|a - a_0| + |b - b_0|) \|v\| . \quad (4.34)$$

Proof (i) The estimate (4.26) gives (4.31). Applying the first inequality in Lemma 4.5.2 with $(a_1, b_1) = (a, b)$, $(a_2, b_2) = (a_0, b_0)$, $u_1 = \theta(w, a, b) + w$, $u_2 = \theta(w, a_0, b_0) + w$ and noting that $u_1 - u_2 = \theta(w, a, b) - \theta(w, a_0, b_0) \perp I'(u_1, a_1, b_1) - I'(u_2, a_2, b_2)$ gives

$$\underline{\lambda} \|\theta(w, a, b) - \theta(w, a_0, b_0)\|^2 - (|a_1 - a_2| + |b_1 - b_2|) \|\theta(w, a_0, b_0)$$
$$+ w\| \|\theta(w, a, b) - \theta(w, a_0, b_0)\|$$
$$\leqslant \|\theta(w, a, b) - \theta(w, a_0, b_0)\|_D^2$$
$$\leqslant \lambda_{l-1} \|\theta(w, a, b) - \theta(w, a_0, b_0)\|^2 \qquad (4.35)$$

where $\underline{\lambda} = \min\{a, b\}$. Since $\underline{\lambda} - \lambda_{l-1}$ is uniformly positive in N, (4.32) follows from (4.31) and (4.35).

(*ii*) The estimate (4.29) gives (4.33). Applying the second inequality in Lemma 4.5.2 with $(a_1, b_1) = (a, b)$, $(a_2, b_2) = (a_0, b_0)$, $u_1 = v + \tau(v, a, b)$, $u_2 = v + \tau(v, a_0, b_0)$ and noting that $u_1 - u_2 = \tau(v, a, b) - \tau(v, a_0, b_0) \perp I'(u_1, a_1, b_1) - I'(u_2, a_2, b_2)$ gives

$$\|\tau(v, a, b) - \tau(v, a_0, b_0)\|_D^2 - (|a_1 - a_2| + |b_1 - b_2|) \|v$$

$$+ \tau(v, a_0, b_0)\| \, \|\tau(v, a, b) - \tau(v, a_0, b_0)\|$$

$$\leqslant \bar{\lambda} \|\tau(v, a, b) - \tau(v, a_0, b_0)\|^2$$

$$\leqslant \bar{\lambda} \|\tau(v, a, b) - \tau(v, a_0, b_0)\|_D^2 / \lambda_{l+1} \tag{4.36}$$

where $\bar{\lambda} = \max\{a, b\}$. Since $1 - \bar{\lambda}/\lambda_{l+1}$ is uniformly positive in N, (4.34) follows from (4.33) and (4.36). $\qquad\square$

Corollary 4.7.3 *We have*

(*i*) θ *is continuous on* $M_{l-1} \times Q_l$,
(*ii*) τ *is continuous on* $N_l \times Q_l$.

Proof (*i*) If $(w_j, a_j, b_j) \to (w, a, b)$ in $M_{l-1} \times Q_l$, then

$$\|\theta(w_j, a_j, b_j) - \theta(w, a, b)\| \leqslant \|\theta(w_j, a_j, b_j) - \theta(w_j, a, b)\|$$

$$+ \|\theta(w_j, a, b) - \theta(w, a, b)\|$$

and the last term goes to zero by the continuity of $\theta(\cdot, a, b)$. By Lemma 4.7.2 (*i*) there is a constant $C > 0$ such that for sufficiently large j,

$$\|\theta(w_j, a_j, b_j) - \theta(w_j, a, b)\| \leqslant C \, (|a_j - a| + |b_j - b|) \, \|w_j\|_D \to 0.$$

(*ii*) If $(v_j, a_j, b_j) \to (v, a, b)$ in $N_l \times Q_l$, then

$$\|\tau(v_j, a_j, b_j) - \tau(v, a, b)\|_D \leqslant \|\tau(v_j, a_j, b_j) - \tau(v_j, a, b)\|_D$$

$$+ \|\tau(v_j, a, b) - \tau(v, a, b)\|_D$$

and the last term goes to zero by the continuity of $\tau(\cdot, a, b)$. By Lemma 4.7.2 (*ii*) there is a constant $C > 0$ such that for sufficiently large j,

$$\|\tau(v_j, a_j, b_j) - \tau(v_j, a, b)\|_D \leqslant C \, (|a_j - a| + |b_j - b|) \, \|v_j\| \to 0. \qquad\square$$

When $a = b = \lambda_l$, we have the following proposition.

Proposition 4.7.4 *We have*

(*i*) $\theta(w, \lambda_l, \lambda_l) = 0 \quad \forall w \in M_{l-1},$

(*ii*) $\tau(v, \lambda_l, \lambda_l) = 0 \quad \forall v \in N_l.$

Proof (*i*) Follows from (4.23) since

$$I'(w, \lambda_l, \lambda_l) = 2(A - \lambda_l) w \perp N_{l-1}.$$

(*ii*) Follows from (4.25) since

$$I'(v, \lambda_l, \lambda_l) = 2(A - \lambda_l) v \perp M_l. \qquad \square$$

For $(a, b) \in Q_l$, let

$$\sigma(w, a, b) = \theta(w, a, b) + w, \quad w \in M_{l-1},$$

$$S_l(a, b) = \sigma(M_{l-1}, a, b),$$

$$\zeta(v, a, b) = v + \tau(v, a, b), \quad v \in N_l,$$

$$S^l(a, b) = \zeta(N_l, a, b).$$

Then S_l, S^l are topological manifolds modeled on M_{l-1}, N_l, respectively. Thus, S_l is infinite dimensional, while S^l is d_l-dimensional, where

$$d_l = \dim N_l.$$

For $B \subset D$, set

$$\tilde{B} = B \cap S, \tag{4.37}$$

where

$$S = \left\{ u \in D : \|u\|_D = 1 \right\}$$

is the unit sphere in D. We say that B is a radial set if

$$B = \left\{ su : u \in \tilde{B}, s \geq 0 \right\}.$$

Since θ, τ are positive homogeneous, so are σ, ζ, and hence S_l, S^l are radial manifolds.

Let

$$K(a, b) = \left\{ u \in D : I'(u, a, b) = 0 \right\}$$

be the set of critical points of $I(\cdot, a, b)$. Since I' is positive homogeneous, K is a radial set. Moreover,

$$I(u) = \frac{1}{2} \left(I'(u), u \right) \tag{4.38}$$

by Proposition 4.3.2 and hence

$$I(u) = 0 \quad \forall u \in K. \tag{4.39}$$

Since

$$D = N_{l-1} \oplus E_l \oplus M_l,$$

Proposition 4.7.1 implies

$$K = \left\{ u \in S_l \cap S^l : I'(u) \perp E_l \right\}. \tag{4.40}$$

Together with (4.39), it also implies

$$K \subset \left\{ u \in S_l \cap S^l : I(u) = 0 \right\}. \tag{4.41}$$

Set

$$n_{l-1}(a, b) = \inf_{w \in \tilde{M}_{l-1}} \sup_{v \in N_{l-1}} I(v + w, a, b),$$

$$m_l(a, b) = \sup_{v \in \tilde{N}_l} \inf_{w \in M_l} I(v + w, a, b).$$

Since $I(u, a, b)$ is nonincreasing in a for fixed u, b and in b for fixed u, a, $n_{l-1}(a, b), m_l(a, b)$ are nonincreasing in a for fixed b and in b for fixed a. By Proposition 4.7.1,

$$n_{l-1}(a, b) = \inf_{w \in \tilde{M}_{l-1}} I(\sigma(w, a, b), a, b), \tag{4.42}$$

$$m_l(a, b) = \sup_{v \in \tilde{N}_l} I(\zeta(v, a, b), a, b). \tag{4.43}$$

Proposition 4.7.5 *Let* $(a, b), (a', b') \in Q_l$.

(i) *Assume that* $n_{l-1}(a, b) = 0$. *Then*

$$I(u, a, b) \geqslant 0 \quad \forall u \in S_l(a, b), \tag{4.44}$$

$$K(a, b) = \left\{ u \in S_l(a, b) : I(u, a, b) = 0 \right\}, \tag{4.45}$$

and $(a, b) \in \Sigma(A)$.
  (a) *If* $a' \leqslant a, b' \leqslant b$, *and* $(a', b') \neq (a, b)$, *then* $n_{l-1}(a', b') > 0$,

$$I(u, a', b') > 0 \quad \forall u \in S_l(a', b') \setminus \{0\}, \tag{4.46}$$

and $(a', b') \notin \Sigma(A)$.
  (b) *If* $a' \geqslant a, b' \geqslant b$, *and* $(a', b') \neq (a, b)$, *then* $n_{l-1}(a', b') < 0$ *and there is a* $u \in S_l(a', b') \setminus \{0\}$ *such that*

$$I(u, a', b') < 0.$$

(ii) *Assume that* $m_l(a, b) = 0$. *Then*

$$I(u, a, b) \leqslant 0 \quad \forall u \in S^l(a, b), \tag{4.47}$$

$$K(a, b) = \left\{ u \in S^l(a, b) : I(u, a, b) = 0 \right\}, \tag{4.48}$$

and $(a, b) \in \Sigma(A)$.
(a) *If* $a' \geqslant a$, $b' \geqslant b$, *and* $(a', b') \neq (a, b)$, *then* $m_l(a', b') < 0$,

$$I(u, a', b') < 0 \quad \forall u \in S^l(a', b') \backslash \{0\}, \tag{4.49}$$

and $(a', b') \notin \Sigma(A)$.
(b) *If* $a' \leqslant a$, $b' \leqslant b$, *and* $(a', b') \neq (a, b)$, *then* $m_l(a', b') > 0$ *and there is a* $u \in S^l(a', b') \backslash \{0\}$ *such that*

$$I(u, a', b') > 0.$$

Proof (i) By (4.42),

$$I(\sigma(w)) \geqslant \|w\|_D^2 \, n_{l-1} = 0 \quad \forall w \in M_{l-1}, \tag{4.50}$$

so (4.44) holds.

By (4.41), K is contained in the set B on the right-hand side of (4.45). Noting that

$$B = \left\{ \sigma(w) : w \in M_{l-1}, \ I(\sigma(w)) = 0 \right\},$$

suppose $\sigma(w) \in B$, so

$$I(\sigma(w)) = 0. \tag{4.51}$$

To show that $I'(\sigma(w)) = 0$ and hence $\sigma(w) \in K$, it suffices to check that

$$\left(I'(\sigma(w)), z \right) = 0 \quad \forall z \in M_{l-1} \tag{4.52}$$

since $I'(\sigma(w)) \perp N_{l-1}$ by Proposition 4.7.1 (i). For $t \in \mathbb{R}$,

$$I(\theta(w + tz) + w + tz) - I(\theta(w + tz) + w) \geqslant I(\sigma(w + tz)) - I(\sigma(w)) \geqslant 0$$

by Proposition 4.7.1 (i), (4.50), and (4.51). So

$$\int_0^1 \frac{d}{ds} \left(I(\theta(w + tz) + w + stz) \right) ds \geqslant 0,$$

or

$$t \int_0^1 \left(I'(\theta(w + tz) + w + stz), z \right) ds \geqslant 0.$$

Dividing by $t > 0$ and letting $t \searrow 0$ gives $(I'(\sigma(w)), z) \geqslant 0$, and dividing by $t < 0$ and letting $t \nearrow 0$ gives $(I'(\sigma(w)), z) \leqslant 0$, so (4.52) holds.

To show that $(a, b) \in \Sigma(A)$, it now suffices to produce a nonzero element of B. Let $(w_j) \subset \tilde{M}_{l-1}$ be a minimizing sequence for n_{l-1} in (4.42), so

$$I(\sigma(w_j)) \to 0.$$

A renamed subsequence converges to some $w \in M_{l-1}$ weakly in D and strongly in H since the embedding $D \hookrightarrow H$ is compact. Then $\theta(w_j) \to \theta(w)$ by (4.28) and hence $\sigma(w_j) \to \sigma(w)$ in D, so (4.51) follows from the weak lower semi-continuity of I. By Proposition 4.7.1 (i),

$$I(\sigma(w_j)) \geqslant I(w_j) = 1 - a \|w_j^-\|^2 - b \|w_j^+\|^2.$$

Passing to the limit gives

$$a \|w^-\|^2 + b \|w^+\|^2 \geqslant 1,$$

so $w \neq 0$ and hence $\sigma(w) \neq 0$.

(a) Since $I(\cdot, a', b') \geqslant I(\cdot, a, b)$,

$$n_{l-1}(a', b') \geqslant n_{l-1}(a, b) = 0.$$

Suppose $n_{l-1}(a', b') = 0$. As above, there is a $w' \in M_{l-1} \backslash \{0\}$ such that

$$I(\sigma(w', a', b'), a', b') = 0.$$

Let $u = \sigma(w', a, b) \in S_l(a, b) \backslash \{0\}$. Then

$$
\begin{aligned}
0 &\leqslant I(u, a, b) && \text{by (4.44)} \\
&\leqslant I(u, a', b') \\
&\leqslant I(\sigma(w', a', b'), a', b') && \text{by Proposition 4.7.1 } (i) \\
&= 0
\end{aligned}
$$

and hence equality holds throughout, so $u \in K(a, b)$ by (4.45). Moreover,

$$0 \leqslant (a - a') \|u^-\|^2 + (b - b') \|u^+\|^2 = I(u, a', b') - I(u, a, b) = 0$$

and hence u^- or u^+ is zero. Thus, either $u = u^+$ is a nontrivial solution of

$$Au = bu,$$

or $u = -u^-$ is a nontrivial solution of

$$Au = au.$$

Since λ_l is the only eigenvalue of A in $(\lambda_{l-1}, \lambda_{l+1})$, $u \in E_l$ in either case. This contradicts (4.7) since $l \geqslant 2$, so $n_{l-1}(a', b') > 0$.

Then

$$I(\sigma(w, a', b'), a', b') \geq \|w\|_D^2 \, n_{l-1}(a', b') > 0 \quad \forall w \in M_{l-1} \setminus \{0\}$$

by (4.42), so (4.46) holds. Hence $K(a', b') = \{0\}$ by (4.41), so $(a', b') \notin \Sigma(A)$.

(b) Since $I(\cdot, a', b') \leq I(\cdot, a, b)$,

$$n_{l-1}(a', b') \leq n_{l-1}(a, b) = 0.$$

If $n_{l-1}(a', b') = 0$, then $n_{l-1}(a, b) > 0$ by (a), so $n_{l-1}(a', b') < 0$. Then there is a $w \in M_{l-1} \setminus \{0\}$ such that

$$I(\sigma(w, a', b'), a', b') < 0$$

by (4.42).

(ii) By (4.43),

$$I(\zeta(v)) \leq \|v\|_D^2 \, m_l = 0 \quad \forall v \in N_l, \tag{4.53}$$

so (4.47) holds.

By (4.41), K is contained in the set C on the right-hand side of (4.48). Noting that

$$C = \{\zeta(v) : v \in N_l, \ I(\zeta(v)) = 0\},$$

suppose $\zeta(v) \in C$, so

$$I(\zeta(v)) = 0. \tag{4.54}$$

To show that $I'(\zeta(v)) = 0$ and hence $\zeta(v) \in K$, it suffices to check that

$$\big(I'(\zeta(v)), y\big) = 0 \quad \forall y \in N_l \tag{4.55}$$

since $I'(\zeta(v)) \perp M_l$ by Proposition 4.7.1 (ii). For $t \in \mathbb{R}$,

$$I(v + ty + \tau(v + ty)) - I(v + \tau(v + ty)) \leq I(\zeta(v + ty)) - I(\zeta(v)) \leq 0$$

by Proposition 4.7.1 (ii), (4.53), and (4.54). So

$$\int_0^1 \frac{d}{ds} \big(I(v + sty + \tau(v + ty))\big) \, ds \leq 0,$$

or

$$t \int_0^1 \big(I'(v + sty + \tau(v + ty)), y\big) \, ds \leq 0.$$

Dividing by $t > 0$ and letting $t \searrow 0$ gives $(I'(\zeta(v)), y) \leq 0$, and dividing by $t < 0$ and letting $t \nearrow 0$ gives $(I'(\zeta(v)), y) \geq 0$, so (4.55) holds.

To show that $(a, b) \in \Sigma(A)$, it now suffices to produce a nonzero element of C. Since N_l is finite dimensional, the supremum in (4.43) is achieved at some $v \in \widetilde{N}_l$, so (4.54) holds and $\zeta(v) \neq 0$.

(a) Since $I(\cdot, a', b') \leqslant I(\cdot, a, b)$,

$$m_l(a', b') \leqslant m_l(a, b) = 0.$$

Suppose $m_l(a', b') = 0$. As above, there is a $v' \in N_l \backslash \{0\}$ such that

$$I(\zeta(v', a', b'), a', b') = 0.$$

Let $u = \zeta(v', a, b) \in S^l(a, b) \backslash \{0\}$. Then

$$
\begin{aligned}
0 \geqslant I(u, a, b) \qquad & \text{by (4.47)} \\
\geqslant I(u, a', b') & \\
\geqslant I(\zeta(v', a', b'), a', b') \quad & \text{by Proposition 4.7.1 (ii)} \\
= 0 &
\end{aligned}
$$

and hence equality holds throughout, so $u \in K(a, b)$ by (4.48). Moreover,

$$0 \leqslant (a' - a) \| u^- \|^2 + (b' - b) \| u^+ \|^2 = I(u, a, b) - I(u, a', b') = 0$$

and hence u^- or u^+ is zero. Thus, either $u = u^+$ is a nontrivial solution of

$$Au = bu,$$

or $u = -u^-$ is a nontrivial solution of

$$Au = au.$$

Since λ_l is the only eigenvalue of A in $(\lambda_{l-1}, \lambda_{l+1})$, $u \in E_l$ in either case. This contradicts (4.7) since $l \geqslant 2$, so $m_l(a', b') < 0$.

Then

$$I(\zeta(v, a', b'), a', b') \leqslant \|v\|_D^2 \, m_l(a', b') < 0 \quad \forall v \in N_l \backslash \{0\}$$

by (4.43), so (4.49) holds. Hence $K(a', b') = \{0\}$ by (4.41), so $(a', b') \notin \Sigma(A)$.

(b) Since $I(\cdot, a', b') \geqslant I(\cdot, a, b)$,

$$m_l(a', b') \geqslant m_l(a, b) = 0.$$

If $m_l(a', b') = 0$, then $m_l(a, b) < 0$ by (a), so $m_l(a', b') > 0$. Then there is a $v \in N_l \backslash \{0\}$ such that

$$I(\zeta(v, a', b'), a', b') > 0$$

by (4.43).   □

Next we show that n_{l-1}, m_l depend continuously on (a, b).

Lemma 4.7.6 *Given* $(a_0, b_0) \in Q_l$ *and a neighborhood* $N \subset\subset Q_l$ *of* (a_0, b_0), *there is a constant* $C > 0$ *such that*

(i) for all $(a, b) \in N$,

$$|n_{l-1}(a, b) - n_{l-1}(a_0, b_0)| \leqslant C\,(|a - a_0| + |b - b_0|), \qquad (4.56)$$

(ii) for all $(a, b) \in N$,

$$|m_l(a, b) - m_l(a_0, b_0)| \leqslant C\,(|a - a_0| + |b - b_0|). \qquad (4.57)$$

Proof (i) By Lemmas 4.5.1 and 4.7.2 (i), there is a constant $C > 0$ such that for all $(a, b) \in N$,

$$|I(\sigma(w, a, b), a, b) - I(\sigma(w, a_0, b_0), a_0, b_0)|$$
$$\leqslant C\,(|a - a_0| + |b - b_0|)\,\|w\|_D^2 \quad \forall w \in M_{l-1},$$

which together with (4.42) gives (4.56).

(ii) By Lemmas 4.5.1 and 4.7.2 (ii), there is a constant $C > 0$ such that for all $(a, b) \in N$,

$$|I(\zeta(v, a, b), a, b) - I(\zeta(v, a_0, b_0), a_0, b_0)|$$
$$\leqslant C\,(|a - a_0| + |b - b_0|)\,\|v\|^2 \quad \forall v \in N_l,$$

which together with (4.43) gives (4.57). $\qquad \square$

When $a = b = \lambda_l$, we have the following proposition.

Proposition 4.7.7 *We have*

(i) $n_{l-1}(\lambda_l, \lambda_l) = 0$,
(ii) $m_l(\lambda_l, \lambda_l) = 0$.

Proof (i) By Proposition 4.7.4 (i), $\sigma(w, \lambda_l, \lambda_l) = w$ for $w \in M_{l-1}$, so (4.42) gives

$$n_{l-1}(\lambda_l, \lambda_l) = \inf_{w \in \widetilde{M}_{l-1}} I(w, \lambda_l, \lambda_l).$$

We have

$$I(w, \lambda_l, \lambda_l) = \|w\|_D^2 - \lambda_l \|w\|^2 \geqslant 0$$

by (4.6) and equality holds for $w \in \widetilde{E}_l$.

(*ii*) By Proposition 4.7.4 (*ii*), $\zeta(v, \lambda_l, \lambda_l) = v$ for $v \in N_l$, so (4.43) gives

$$m_l(\lambda_l, \lambda_l) = \sup_{v \in \tilde{N}_l} I(v, \lambda_l, \lambda_l).$$

We have

$$I(v, \lambda_l, \lambda_l) = \|v\|_D^2 - \lambda_l \|v\|^2 \leqslant 0$$

by (4.5) and equality holds for $v \in \tilde{E}_l$. $\qquad\square$

For $a \in (\lambda_{l-1}, \lambda_{l+1})$, set

$$v_{l-1}(a) = \sup \{b \in (\lambda_{l-1}, \lambda_{l+1}) : n_{l-1}(a, b) \geqslant 0\},$$

$$\mu_l(a) = \inf \{b \in (\lambda_{l-1}, \lambda_{l+1}) : m_l(a, b) \leqslant 0\}.$$

Lemma 4.7.8 *Let* $(a, b) \in Q_l$.

(*i*) $b = v_{l-1}(a) \iff n_{l-1}(a, b) = 0.$
(*ii*) $b = \mu_l(a) \iff m_l(a, b) = 0.$

Proof (*i*) Since $n_{l-1}(a, \cdot)$ is continuous on $(\lambda_{l-1}, \lambda_{l+1})$ by Lemma 4.7.6 (*i*), forward implication holds. Reverse implication follows from Proposition 4.7.5 (*i*) (*b*).

(*ii*) Since $m_l(a, \cdot)$ is continuous on $(\lambda_{l-1}, \lambda_{l+1})$ by Lemma 4.7.6 (*ii*), forward implication holds. Reverse implication follows from Proposition 4.7.5 (*ii*) (*b*). $\qquad\square$

The main theorem of this section is as follows.

Theorem 4.7.9 *Let* $(a, b) \in Q_l$.

(*i*) *The function* v_{l-1} *is continuous, strictly decreasing, and satisfies*
  (*a*) $v_{l-1}(\lambda_l) = \lambda_l$,
  (*b*) $b = v_{l-1}(a) \implies (a, b) \in \Sigma(A)$,
  (*c*) $b < v_{l-1}(a) \implies (a, b) \notin \Sigma(A)$.
(*ii*) *The function* μ_l *is continuous, strictly decreasing, and satisfies*
  (*a*) $\mu_l(\lambda_l) = \lambda_l$,
  (*b*) $b = \mu_l(a) \implies (a, b) \in \Sigma(A)$,
  (*c*) $b > \mu_l(a) \implies (a, b) \notin \Sigma(A)$.
(*iii*) $v_{l-1}(a) \leqslant \mu_l(a)$.

Proof (*i*) To see that v_{l-1} is continuous, let $(a_0, v_{l-1}(a_0)) \in Q_l$. Given $\varepsilon > 0$, let $0 < \tilde{\varepsilon} \leqslant \varepsilon$ be so small that $v_{l-1}(a_0) \pm \tilde{\varepsilon} \in (\lambda_{l-1}, \lambda_{l+1})$. Since $n_{l-1}(a_0, v_{l-1}(a_0)) = 0$ by Lemma 4.7.8 (*i*),

$$n_{l-1}(a_0, v_{l-1}(a_0) \pm \tilde{\varepsilon}) \lessgtr 0$$

by Proposition 4.7.5 (*i*). Then there is a $\delta > 0$ such that

$$|a - a_0| < \delta \implies n_{l-1}(a, v_{l-1}(a_0) \pm \tilde{\varepsilon}) \lessgtr 0$$

since $n_{l-1}(\cdot, v_{l-1}(a_0) \pm \tilde{\varepsilon})$ are continuous by Lemma 4.7.6 (*i*). Since $n_{l-1}(a, \cdot)$ is nonincreasing and $n_{l-1}(a, v_{l-1}(a)) = 0$, then

$$|a - a_0| < \delta \implies |v_{l-1}(a) - v_{l-1}(a_0)| < \tilde{\varepsilon}.$$

To see that v_{l-1} is strictly decreasing, let $(a_1, v_{l-1}(a_1)), (a_2, v_{l-1}(a_2)) \in Q_l$ with $a_1 < a_2$. Since $n_{l-1}(a_1, v_{l-1}(a_1)) = 0$ by Lemma 4.7.8 (*i*),

$$n_{l-1}(a_2, v_{l-1}(a_1)) < 0$$

by Proposition 4.7.5 (*i*) (*b*). Then

$$v_{l-1}(a_2) < v_{l-1}(a_1)$$

since $n_{l-1}(a_2, \cdot)$ is nonincreasing and $n_{l-1}(a_2, v_{l-1}(a_2)) = 0$.

(*a*) Follows from Proposition 4.7.7 (*i*) and Lemma 4.7.8 (*i*).

(*b*) We have $n_{l-1}(a, b) = 0$ by Lemma 4.7.8 (*i*), so $(a, b) \in \Sigma(A)$ by Proposition 4.7.5 (*i*).

(*c*) We have $n_{l-1}(a, b) \geq 0$, and since $b < v_{l-1}(a)$, $n_{l-1}(a, b) > 0$ by Lemma 4.7.8 (*i*). Then $(a, b) \notin \Sigma(A)$ as in the proof of Proposition 4.7.5 (*i*) (*a*).

(*ii*) To see that μ_l is continuous, let $(a_0, \mu_l(a_0)) \in Q_l$. Given $\varepsilon > 0$, let $0 < \tilde{\varepsilon} \leq \varepsilon$ be so small that $\mu_l(a_0) \pm \tilde{\varepsilon} \in (\lambda_{l-1}, \lambda_{l+1})$. Since $m_l(a_0, \mu_l(a_0)) = 0$ by Lemma 4.7.8 (*ii*),

$$m_l(a_0, \mu_l(a_0) \pm \tilde{\varepsilon}) \lessgtr 0$$

by Proposition 4.7.5 (*ii*). Then there is a $\delta > 0$ such that

$$|a - a_0| < \delta \implies m_l(a, \mu_l(a_0) \pm \tilde{\varepsilon}) \lessgtr 0$$

since $m_l(\cdot, \mu_l(a_0) \pm \tilde{\varepsilon})$ are continuous by Lemma 4.7.6 (*ii*). Since $m_l(a, \cdot)$ is nonincreasing and $m_l(a, \mu_l(a)) = 0$, then

$$|a - a_0| < \delta \implies |\mu_l(a) - \mu_l(a_0)| < \tilde{\varepsilon}.$$

To see that μ_l is strictly decreasing, let $(a_1, \mu_l(a_1)), (a_2, \mu_l(a_2)) \in Q_l$ with $a_1 < a_2$. Since $m_l(a_1, \mu_l(a_1)) = 0$ by Lemma 4.7.8 (*ii*),

$$m_l(a_2, \mu_l(a_1)) < 0$$

by Proposition 4.7.5 (*ii*) (*a*). Then

$$\mu_l(a_2) < \mu_l(a_1)$$

since $m_l(a_2, \cdot)$ is nonincreasing and $m_l(a_2, \mu_l(a_2)) = 0$.

(a) Follows from Proposition 4.7.7 (ii) and Lemma 4.7.8 (ii).

(b) We have $m_l(a, b) = 0$ by Lemma 4.7.8 (ii), so $(a, b) \in \Sigma(A)$ by Proposition 4.7.5 (ii).

(c) We have $m_l(a, b) \leqslant 0$, and since $b > \mu_l(a)$, $m_l(a, b) < 0$ by Lemma 4.7.8 (ii). Then $(a, b) \notin \Sigma(A)$ as in the proof of Proposition 4.7.5 (ii) (a).

(iii) By (i) (a) and (ii) (a),

$$v_{l-1}(\lambda_l) = \lambda_l = \mu_l(\lambda_l).$$

If $a \in (\lambda_{l-1}, \lambda_l)$ and $v_{l-1}(a) > \mu_l(a)$, then

$$\lambda_l < \mu_l(a) < v_{l-1}(a) \leqslant \lambda_{l+1}$$

since μ_l is strictly decreasing, so $(a, \mu_l(a)) \in \Sigma(A)$ by (ii) (b), contradicting (i) (c). If $a \in (\lambda_l, \lambda_{l+1})$ and $v_{l-1}(a) > \mu_l(a)$, then

$$\lambda_{l-1} \leqslant \mu_l(a) < v_{l-1}(a) < \lambda_l$$

since v_{l-1} is strictly decreasing, so $(a, v_{l-1}(a)) \in \Sigma(A)$ by (i) (b), contradicting (ii) (c). $\square$

Thus,

$$C_l : b = v_{l-1}(a), \qquad C^l : b = \mu_l(a)$$

are strictly decreasing curves in Q_l that belong to $\Sigma(A)$. They both pass through the point (λ_l, λ_l) and may coincide. The region

$$I_l = \{(a, b) \in Q_l : b < v_{l-1}(a)\}$$

below the lower curve C_l and the region

$$I^l = \{(a, b) \in Q_l : b > \mu_l(a)\}$$

above the upper curve C^l are free of $\Sigma(A)$. They are the minimal and maximal curves of $\Sigma(A)$ in Q_l in this sense. Points in the region

$$II_l = \{(a, b) \in Q_l : v_{l-1}(a) < b < \mu_l(a)\}$$

between C_l, C^l, when it is nonempty, may or may not belong to $\Sigma(A)$.

4.8 Null manifold

For $(a, b) \in Q_l$, let

$$\mathcal{N}_l(a, b) = S_l(a, b) \cap S^l(a, b)$$

(see p. 84). Since S_l, S^l are radial sets, so is $\mathcal{N}_l$. We will see that $\mathcal{N}_l$ is a topological manifold modeled on E_l and hence

$$\dim \mathcal{N}_l = d_l - d_{l-1}.$$

We will call it the null manifold of I.

Proposition 4.8.1 *Let* $(a, b) \in Q_l$.

(i) *There is a positive homogeneous map* $\eta(\cdot, a, b) \in C(E_l, N_{l-1})$ *such that* $v = \eta(y)$ *is the unique solution of*

$$I(\zeta(v + y)) = \sup_{v' \in N_{l-1}} I(\zeta(v' + y)), \quad y \in E_l. \tag{4.58}$$

Moreover,

$$I'(\zeta(v + y)) \perp N_{l-1} \iff v = \eta(y). \tag{4.59}$$

(ii) *There is a positive homogeneous map* $\xi(\cdot, a, b) \in C(E_l, M_l)$ *such that* $w = \xi(y)$ *is the unique solution of*

$$I(\sigma(y + w)) = \inf_{w' \in M_l} I(\sigma(y + w')), \quad y \in E_l. \tag{4.60}$$

Moreover,

$$I'(\sigma(y + w)) \perp M_l \iff w = \xi(y). \tag{4.61}$$

(iii) *For all* $y \in E_l$,

$$\zeta(\eta(y) + y) = \sigma(y + \xi(y)), \tag{4.62}$$

i.e.,

$$\eta(y) = \theta(y + \xi(y)), \qquad \xi(y) = \tau(\eta(y) + y). \tag{4.63}$$

First we prove a lemma.

Lemma 4.8.2 *If*

$$u_i = v_i + y + w_i \in N_{l-1} \oplus E_l \oplus M_l, \quad I'(u_i) \perp E_l^{\perp}, \quad i = 1, 2,$$

then $u_1 = u_2$.

Proof By Proposition 4.7.1,

$$u_1 = \sigma(y + w_1)$$

since $I'(u_1) \perp N_{l-1}$ and

$$u_2 = \zeta(v_2 + y)$$

since $I'(u_2) \perp M_l$, so

$$I(u_1) \geqslant I(v_2 + y + w_1) \geqslant I(u_2).$$

First inequality is strict if $v_1 \neq v_2$ and the second is strict if $w_1 \neq w_2$, so

$$I(u_1) > I(u_2)$$

if $u_1 \neq u_2$. This is impossible since interchanging u_1 and u_2 then gives the reverse inequality. □

Proof of Proposition 4.8.1 (i) For $v \in N_{l-1}$, $y \in E_l$,

$$
\begin{aligned}
I(\zeta(v+y)) &\leqslant I(v+y) &&\text{by Prop. 4.7.1 (ii)}\\
&\leqslant \|v+y\|_D^2 - \underline{\lambda}\|v+y\|^2 &&\text{by (4.19)}\\
&\leqslant -(\underline{\lambda}-\lambda_{l-1})\|v\|^2 + \|y\|_D^2 - \underline{\lambda}\|y\|^2 &&\text{by (4.5).}
\end{aligned}
$$
(4.64)

Since $\underline{\lambda} > \lambda_{l-1}$, this implies that $I(\zeta(\cdot + y))$ is bounded from above and anti-coercive on the finite-dimensional space N_{l-1}, so (4.58) has a solution $v = \eta(y)$ satisfying

$$I'(\zeta(\eta(y) + y)) \perp N_{l-1}.$$

Since any solution v of (4.58) satisfies $I'(\zeta(v+y)) \perp N_{l-1}$, to prove uniqueness and (4.59) it only remains to show that

$$I'(\zeta(v_i + y)) \perp N_{l-1}, \ i = 1, 2 \implies v_1 = v_2.$$

Apply Lemma 4.8.2 with $u_i = v_i + y + \tau(v_i + y)$ and note that $I'(u_i) \perp M_l$ by Proposition 4.7.1 (ii).

For $s \geqslant 0$, by (4.59),

$$I'(\zeta(s\,\eta(y) + sy)) = s\,I'(\zeta(\eta(y) + y)) \perp N_{l-1}$$

and hence

$$\eta(sy) = s\,\eta(y).$$

(ii) For $y \in E_l$, $w \in M_l$,

$$
\begin{aligned}
I(\sigma(y+w)) &\geqslant I(y+w) &&\text{by Prop. 4.7.1 (i)}\\
&\geqslant \|y+w\|_D^2 - \bar{\lambda}\|y+w\|^2 &&\text{by (4.19)}\\
&\geqslant (1 - \bar{\lambda}/\lambda_{l+1})\|w\|_D^2 + \|y\|_D^2 - \bar{\lambda}\|y\|^2 &&\text{by (4.6).}
\end{aligned}
$$

Since $\bar{\lambda} < \lambda_{l+1}$, this implies that $I(\sigma(y + \cdot))$ is bounded from below and coercive on M_l. It is also weakly lower semicontinuous since the embedding

$D \hookrightarrow H$ is compact. So (4.60) has a solution $w = \xi(y)$ satisfying

$$I'(\sigma(y + \xi(y))) \perp M_l.$$

Since any solution w of (4.60) satisfies $I'(\sigma(y + w)) \perp M_l$, to prove unique-ness and (4.61) it only remains to show that

$$I'(\sigma(y + w_i)) \perp M_l, \ i = 1, 2 \implies w_1 = w_2.$$

Apply Lemma 4.8.2 with $u_i = \theta(y + w_i) + y + w_i$ and note that $I'(u_i) \perp N_{l-1}$ by Proposition 4.7.1 (i).

For $s \geqslant 0$, by (4.61),

$$I'(\sigma(sy + s\,\xi(y))) = s\,I'(\sigma(y + \xi(y))) \perp M_l$$

and hence

$$\xi(sy) = s\,\xi(y).$$

(iii) Applying Lemma 4.8.2 with $u_1 = \eta(y) + y + \tau(\eta(y) + y)$, $u_2 = \theta(y + \xi(y)) + y + \xi(y)$ and noting that $I'(u_1) \perp M_l$, $I'(u_2) \perp N_{l-1}$ by Proposition 4.7.1 gives (4.62).

Continuity of η, ξ follows from our next lemma. $\square$

Lemma 4.8.3 *We have*

(i) η *is continuous on* $E_l \times Q_l$,
(ii) ξ *is continuous on* $E_l \times Q_l$.

Proof (i) Let $(y_j, a_j, b_j) \to (y, a, b)$ in $E_l \times Q_l$ and suppose that $v_j = \eta(y_j, a_j, b_j) \nrightarrow \eta(y, a, b)$, so

$$\inf_j \|v_j - \eta(y, a, b)\| > 0 \qquad (4.65)$$

for a renamed subsequence. Since

$$I(\zeta(v_j + y_j, a_j, b_j), a_j, b_j) \geqslant I(\zeta(y_j, a_j, b_j), a_j, b_j),$$

(4.64) gives

$$\|v_j\|^2 \leqslant (\|y_j\|_D^2 - \underline{\lambda}_j \|y_j\|^2 - I(\zeta(y_j, a_j, b_j), a_j, b_j))/(\underline{\lambda}_j - \lambda_{l-1})$$

$$\to (\|y\|_D^2 - \underline{\lambda} \|y\|^2 - I(\zeta(y, a, b), a, b))/(\underline{\lambda} - \lambda_{l-1})$$

where $\underline{\lambda}_j = \min\{a_j, b_j\}$, $\underline{\lambda} = \min\{a, b\}$. So (v_j) is bounded and hence con-verges to some $v \in N_{l-1}$ for a renamed subsequence since N_{l-1} is finite dimen-sional. Then $\zeta(v_j + y_j, a_j, b_j) \to \zeta(v + y, a, b)$ by Corollary 4.7.3 (ii) and hence

$$I'(\zeta(v_j + y_j, a_j, b_j), a_j, b_j) \to I'(\zeta(v + y, a, b), a, b).$$

Since $I'(\zeta(v_j + y_j, a_j, b_j), a_j, b_j) \perp N_{l-1}$, then $I'(\zeta(v + y, a, b), a, b) \perp N_{l-1}$, but $v \neq \eta(y, a, b)$ by (4.65). This contradicts (4.59).

(*ii*) If $(y_j, a_j, b_j) \to (y, a, b)$ in $E_l \times Q_l$, then $\eta(y_j, a_j, b_j) \to \eta(y, a, b)$ by (*i*) and hence

$$\xi(y_j, a_j, b_j) = \tau(\eta(y_j, a_j, b_j) + y_j, a_j, b_j) \to \tau(\eta(y, a, b) + y, a, b)$$
$$= \xi(y, a, b)$$

by (4.63) and Corollary 4.7.3 (*ii*). □

When $a = b = \lambda_l$, we have the following proposition.

Proposition 4.8.4 *We have*

(*i*) $\eta(y, \lambda_l, \lambda_l) = 0 \quad \forall y \in E_l,$
(*ii*) $\xi(y, \lambda_l, \lambda_l) = 0 \quad \forall y \in E_l.$

Proof (*i*) Follows from (4.59) since Proposition 4.7.4 (*ii*) gives

$$I'(\zeta(y, \lambda_l, \lambda_l), \lambda_l, \lambda_l) = I'(y, \lambda_l, \lambda_l) = 2(A - \lambda_l) y = 0.$$

(*ii*) Follows from (4.61) since Proposition 4.7.4 (*i*) gives

$$I'(\sigma(y, \lambda_l, \lambda_l), \lambda_l, \lambda_l) = I'(y, \lambda_l, \lambda_l) = 2(A - \lambda_l) y = 0. \qquad □$$

Referring to Proposition 4.8.1 (*iii*), let

$$\varphi(y) = \zeta(\eta(y) + y) = \sigma(y + \xi(y)), \quad y \in E_l.$$

Proposition 4.8.5 *Let* $(a, b) \in Q_l.$

(*i*) $\varphi(\cdot, a, b) \in C(E_l, D)$ *is a positive homogeneous map such that*

$$I(\varphi(y)) = \inf_{w \in M_l} \sup_{v \in N_{l-1}} I(v + y + w) = \sup_{v \in N_{l-1}} \inf_{w \in M_l} I(v + y + w), \quad y \in E_l$$

and

$$I'(\varphi(y)) \in E_l \quad \forall y \in E_l.$$

(*ii*) *If* $(a', b') \in Q_l$ *with* $a' \geqslant a$ *and* $b' \geqslant b$, *then*

$$I(\varphi(y, a', b'), a', b') \leqslant I(\varphi(y, a, b), a, b) \quad \forall y \in E_l.$$

(*iii*) φ *is continuous on* $E_l \times Q_l.$
(*iv*) $\varphi(y, \lambda_l, \lambda_l) = y \quad \forall y \in E_l.$
(*v*) $N_l(a, b) = \{\varphi(y, a, b) : y \in E_l\}.$
(*vi*) $N_l(\lambda_l, \lambda_l) = E_l.$

Proof (*i*) Follows from Propositions 4.7.1 and 4.8.1 (*i*) and (*ii*).

(*ii*) Follows from (*i*) since $I(\cdot, a', b') \leqslant I(\cdot, a, b).$

(*iii*) Follows from Corollary 4.7.3 and Lemma 4.8.3.

(*iv*) Follows from Propositions 4.7.4 and 4.8.4.

(*v*) Clearly, $\varphi(y)$ is in S_l, S^l and hence in $\mathcal{N}_l$ for each $y \in E_l$. Conversely, let $u \in \mathcal{N}_l$, and write $u = v + y + w \in N_{l-1} \oplus E_l \oplus M_l$. Since $u \in S_l$, S^l,

$$u = \sigma(y + w) = \zeta(v + y).$$

Then $I'(\sigma(y + w)) = I'(\zeta(v + y)) \perp M_l$ by Proposition 4.7.1 (*ii*) and hence

$$w = \xi(y)$$

by Proposition 4.8.1 (*ii*). So

$$u = \sigma(y + \xi(y)) = \varphi(y).$$

(*vi*) Follows from (*iv*) and (*v*). $\hspace{2cm}$ □

By (4.40) and (4.41),

$$K = \{u \in \mathcal{N}_l : I'(u) \perp E_l\} \subset \{u \in \mathcal{N}_l : I(u) = 0\}. \qquad (4.66)$$

The following theorem shows that the curves C_l, C^l are closely related to

$$\tilde{I} = I|_{\mathcal{N}_l}.$$

Theorem 4.8.6 *Let* $(a, b) \in Q_l$.

(*i*) *If* $b < \nu_{l-1}(a)$, *then*

$$\tilde{I}(u, a, b) > 0 \quad \forall u \in \mathcal{N}_l(a, b) \backslash \{0\}.$$

(*ii*) *If* $b = \nu_{l-1}(a)$, *then*

$$\tilde{I}(u, a, b) \geqslant 0 \quad \forall u \in \mathcal{N}_l(a, b),$$

$$K(a, b) = \{u \in \mathcal{N}_l(a, b) : \tilde{I}(u, a, b) = 0\}.$$

(*iii*) *If* $\nu_{l-1}(a) < b < \mu_l(a)$, *then there are* $u_i \in \mathcal{N}_l(a, b) \backslash \{0\}$, $i = 1, 2$ *such that*

$$\tilde{I}(u_1, a, b) < 0 < \tilde{I}(u_2, a, b).$$

(*iv*) *If* $b = \mu_l(a)$, *then*

$$\tilde{I}(u, a, b) \leqslant 0 \quad \forall u \in \mathcal{N}_l(a, b),$$

$$K(a, b) = \{u \in \mathcal{N}_l(a, b) : \tilde{I}(u, a, b) = 0\}.$$

(*v*) *If* $b > \mu_l(a)$, *then*

$$\tilde{I}(u, a, b) < 0 \quad \forall u \in \mathcal{N}_l(a, b) \backslash \{0\}.$$

Proof (*i*) As in the proof of Theorem 4.7.9 (*i*) (*c*), $n_{l-1}(a, b) > 0$. Then $I > 0$ on $S_l \setminus \{0\} \supset \mathcal{N}_l \setminus \{0\}$ as in the proof of Proposition 4.7.5 (*i*) (*a*).

(*ii*) By Lemma 4.7.8 (*i*), $n_{l-1}(a, b) = 0$. Apply Proposition 4.7.5 (*i*) and note that $S_l \supset \mathcal{N}_l \supset K$.

(*iii*) We have $n_{l-1}(a, b) < 0 < m_l(a, b)$. Then there are $\sigma(y_1 + w) \in S_l \setminus \{0\}$, $\zeta(v + y_2) \in S^l \setminus \{0\}$ such that

$$I(\sigma(y_1 + w)) < 0 < I(\zeta(v + y_2))$$

as in the proof of Proposition 4.7.5 (*i*) (*b*) and (*ii*) (*b*). Let $u_1 = \sigma(y_1 + \xi(y_1))$, $u_2 = \zeta(\eta(y_2) + y_2)$. By Proposition 4.8.1 (*i*) and (*ii*), $u_i \in \mathcal{N}_l$ and

$$I(u_1) \leqslant I(\sigma(y_1 + w)), \qquad I(u_2) \geqslant I(\zeta(v + y_2)).$$

(*iv*) By Lemma 4.7.8 (*ii*), $m_l(a, b) = 0$. Apply Proposition 4.7.5 (*ii*) and note that $S^l \supset \mathcal{N}_l \supset K$.

(*v*) As in the proof of Theorem 4.7.9 (*ii*) (*c*), $m_l(a, b) < 0$. Then $I < 0$ on $S^l \setminus \{0\} \supset \mathcal{N}_l \setminus \{0\}$ as in the proof of Proposition 4.7.5 (*ii*) (*a*). □

By (4.66), solutions of (4.10) are in $\mathcal{N}_l$. Next we show that the set of solutions is all of $\mathcal{N}_l$ exactly when (a, b) is on both C_l and C^l.

Theorem 4.8.7 *If* $(a, b) \in Q_l$, *then* $K(a, b) = \mathcal{N}_l(a, b)$ *if and only if* $(a, b) \in C_l \cap C^l$.

Proof If $(a, b) \in C_l \cap C^l$, then $\tilde{I} \geqslant 0$ by Theorem 4.8.6 (*ii*) and $\tilde{I} \leqslant 0$ by (*iv*), so $\tilde{I} = 0$ and hence $K = \mathcal{N}_l$ by (*ii*) or (*iv*).

If $K = \mathcal{N}_l$, then $\tilde{I} = 0$ by (4.66) and hence $(a, b) \in C_l \cup C^l$ by Theorem 4.8.6 (*i*), (*iii*), and (*v*). If $(a, b) \in C_l$, then $\tilde{I}(\cdot, a, b') \leqslant \tilde{I}(\cdot, a, b) = 0$ for all $b' \in [b, \lambda_{l+1})$ by Proposition 4.8.5 (*ii*) and hence $(a, b) \in C^l$ by Theorem 4.8.6 (*iii*). If $(a, b) \in C^l$, then $\tilde{I}(\cdot, a, b') \geqslant \tilde{I}(\cdot, a, b) = 0$ for all $b' \in (\lambda_{l-1}, b]$ by Proposition 4.8.5 (*ii*) and hence $(a, b) \in C_l$ by Theorem 4.8.6 (*iii*). □

When λ_l is a simple eigenvalue, $\mathcal{N}_l$ is one-dimensional and hence Theorem 4.8.7 implies the following.

Corollary 4.8.8 *If* λ_l *is simple, then* $(a, b) \in Q_l$ *is on exactly one of the curves* C_l, C^l *if and only if*

$$K(a, b) = \{t\,\varphi(y_0, a, b) : t \geqslant 0\}$$

for some $y_0 \in E_l \setminus \{0\}$.

4.9 Type II regions

In this section we give a sufficient condition for the region II_l to be nonempty (see p. 93).

Theorem 4.9.1 *If there are* $y_i \in E_l$, $i = 1, 2$ *such that*

$$\|y_1^+\| - \|y_1^-\| < 0 < \|y_2^+\| - \|y_2^-\|,$$

then there is a neighborhood $N \subset Q_l$ *of* (λ_l, λ_l) *such that every point* $(a, b) \in N \setminus \{(\lambda_l, \lambda_l)\}$ *with* $a + b = 2\lambda_l$ *is in* II_l.

Proof It suffices to show that there are $u_i \in \mathcal{N}_l(a, b)$, $i = 1, 2$ such that

$$\widetilde{I}(u_1, a, b) < 0 < \widetilde{I}(u_2, a, b) \qquad (4.67)$$

by Theorem 4.8.6 (*i*), (*ii*), (*iv*), and (*v*). Since

$$y_i = \eta(y_i, \lambda_l, \lambda_l) + y_i = y_i + \xi(y_i, \lambda_l, \lambda_l)$$

by Proposition 4.8.4 and $\eta(y_i, \cdot, \cdot) + y_i$, $y_i + \xi(y_i, \cdot, \cdot)$ are continuous on Q_l by Lemma 4.8.3, there is a neighborhood $N \subset Q_l$ of (λ_l, λ_l) such that for $(a, b) \in N$, setting

$$v_i = \eta(y_i, a, b) + y_i, \qquad w_i = y_i + \xi(y_i, a, b)$$

we have

$$\|v_1^+\| - \|v_1^-\| < 0 < \|v_2^+\| - \|v_2^-\|, \qquad \|w_1^+\| - \|w_1^-\| < 0 < \|w_2^+\| - \|w_2^-\|.$$

By Proposition 4.8.5 (*v*), $\varphi(y_i) = \zeta(v_i) = \sigma(w_i) \in \mathcal{N}_l$, and if $a + b = 2\lambda_l$, then

$$\begin{aligned}
\widetilde{I}(\zeta(v_i)) &\leqslant I(v_i) && \text{by Proposition 4.7.1 (\textit{ii})} \\
&\leqslant \lambda_l (\|v_i^+\|^2 + \|v_i^-\|^2) - a\|v_i^-\|^2 \\
&\quad - (2\lambda_l - a)\|v_i^+\|^2 && \text{by (4.5)} \\
&= (a - \lambda_l)(\|v_i^+\|^2 - \|v_i^-\|^2)
\end{aligned}$$

and

$$\begin{aligned}
\widetilde{I}(\sigma(w_i)) &\geqslant I(w_i) && \text{by Proposition 4.7.1 (\textit{i})} \\
&\geqslant \lambda_l (\|w_i^+\|^2 + \|w_i^-\|^2) - a\|w_i^-\|^2 \\
&\quad - (2\lambda_l - a)\|w_i^+\|^2 && \text{by (4.6)} \\
&= (a - \lambda_l)(\|w_i^+\|^2 - \|w_i^-\|^2).
\end{aligned}$$

If $(a, b) \neq (\lambda_l, \lambda_l)$, then $a \neq \lambda_l$ since $a + b = 2\lambda_l$. Thus, (4.67) holds for

$$u_1 = \begin{cases} \varphi(y_2), & a < \lambda_l \\ \varphi(y_1), & a > \lambda_l, \end{cases} \qquad u_2 = \begin{cases} \varphi(y_1), & a < \lambda_l \\ \varphi(y_2), & a > \lambda_l. \end{cases} \qquad \square$$

Corollary 4.9.2 *If* (4.15) *holds and there is a* $y \in E_l$ *such that*

$$\|y^+\| \neq \|y^-\|,$$

then there is a neighborhood $N \subset Q_l$ *of* (λ_l, λ_l) *such that every point* $(a, b) \in N \setminus \{(\lambda_l, \lambda_l)\}$ *with* $a + b = 2\lambda_l$ *is in* II_l.

Proof By (4.15),

$$\|(-y)^+\| - \|(-y)^-\| = -(\|y^+\| - \|y^-\|)$$

and hence $\|(\pm y)^+\| - \|(\pm y)^-\|$ have opposite signs. $\square$

Remark 4.9.3 For problem (4.11), Corollary 4.9.2 is due to Li *et al.* [69].

4.10 Simple eigenvalues

In this section we show that when λ_l is a simple eigenvalue, the region II_l is free of $\Sigma(A)$.

Theorem 4.10.1 *If* λ_l *is simple, then* $\mathrm{II}_l \cap \Sigma(A) = \emptyset$.

Proof If $(a, b) \in \mathrm{II}_l$, then there are $u_i \in \mathcal{N}_l \setminus \{0\}$, $i = 1, 2$ such that

$$\tilde{I}(u_1) < 0 < \tilde{I}(u_2)$$

by Theorem 4.8.6 (*iii*). Since $\mathcal{N}_l$ is one-dimensional and $\tilde{I}$ is positive homogeneous of degree 2, then $\tilde{I} \neq 0$ on $\mathcal{N}_l \setminus \{0\}$ and hence $K = \{0\}$ by (4.66), so $(a, b) \notin \Sigma(A)$. $\square$

Remark 4.10.2 For problem (4.11), Theorem 4.10.1 is due to Gallouët and Kavian [54].

4.11 Critical groups

When $(a, b) \notin \Sigma(A)$, the origin is the only critical point of I, so the critical groups $C_q(I, 0)$ are defined. First we show that they are constant in connected components of $\mathbb{R}^2 \setminus \Sigma(A)$.

Proposition 4.11.1 *If (a_0, b_0) and (a_1, b_1) belong to the same connected component of $\mathbb{R}^2 \backslash \Sigma(A)$, then*

$$C_q(I(\cdot, a_0, b_0), 0) \approx C_q(I(\cdot, a_1, b_1), 0) \quad \forall q.$$

Proof Since $\mathbb{R}^2 \backslash \Sigma(A)$ is open by Proposition 4.4.3, so are its connected components, which are then path connected. Let $[0, 1] \to \mathbb{R}^2 \backslash \Sigma(A)$, $t \mapsto (a_t, b_t)$ be a path joining (a_0, b_0) and (a_1, b_1). Since $(a_t, b_t) \notin \Sigma(A)$, the origin is the only critical point of $I(\cdot, a_t, b_t)$, which satisfies (PS) by Proposition 4.4.2. We apply Theorem 1.4.4 in a closed and bounded neighborhood U of the origin. Clearly, (i) holds. For $t, t_0 \in [0, 1]$, $u \in U$,

$$\left| I(u, a_t, b_t) - I(u, a_{t_0}, b_{t_0}) \right| + \left\| I'(u, a_t, b_t) - I'(u, a_{t_0}, b_{t_0}) \right\|_D$$

$$= \left| (a_t - a_{t_0}) \|u^-\|^2 + (b_t - b_{t_0}) \|u^+\|^2 \right| + 2 \left\| (a_t - a_{t_0}) u^- - (b_t - b_{t_0}) u^+ \right\|_D$$

$$\leqslant C \left(|a_t - a_{t_0}| + |b_t - b_{t_0}| \right)$$

for some constant $C > 0$, so the continuity of a_t, b_t gives (ii). $\qquad\square$

For $B \subset D$, set

$$B^- = \left\{ u \in B : I(u) < 0 \right\}, \qquad B^+ = B \backslash B^-. \tag{4.68}$$

The main theorem of this section is as follows.

Theorem 4.11.2 *Let $(a, b) \in Q_l \backslash \Sigma(A)$.*

(i) *If $(a, b) \in I_l$, then*

$$C_q(I, 0) \approx \delta_{qd_{l-1}} \mathbb{Z}_2.$$

(ii) *If $(a, b) \in I^l$, then*

$$C_q(I, 0) \approx \delta_{qd_l} \mathbb{Z}_2.$$

(iii) *If $(a, b) \in II_l$, then*

$$C_q(I, 0) = 0, \quad q \leqslant d_{l-1} \text{ or } q \geqslant d_l$$

and

$$C_q(I, 0) \approx \widetilde{H}_{q-d_{l-1}-1}(\widetilde{\mathcal{N}}_l^-), \quad d_{l-1} < q < d_l.$$

In particular, $C_q(I, 0) = 0$ for all q when λ_l is simple.

Proof By Proposition 4.4.2, I satisfies (PS) since $(a, b) \notin \Sigma(A)$, so applying Proposition 1.4.1 with $a < 0 = I(0)$, $b = +\infty$ gives

$$C_q(I, 0) \approx H_q(D, I^a). \tag{4.69}$$

Since D is contractible, Lemma 1.4.6 (*ii*) gives

$$H_q(D, I^a) \approx \tilde{H}_{q-1}(I^a). \tag{4.70}$$

By Remark 1.3.8, I^a is a strong deformation retract of D^- and hence

$$\tilde{H}_{q-1}(I^a) \approx \tilde{H}_{q-1}(D^-). \tag{4.71}$$

Writing $u \in D^-$ as $v + w \in N_l \oplus M_l$, let

$$\eta_1(u, t) = v + (1 - t) w + t \tau(v), \quad (u, t) \in D^- \times [0, 1].$$

We have

$$\begin{aligned} I(\eta_1(u, t)) &\leqslant (1 - t) I(v + w) + t I(v + \tau(v)) && \text{by Proposition 4.6.1 (\emph{ii})} \\ &\leqslant I(u) && \text{by Proposition 4.7.1 (\emph{ii})} \\ &< 0, \end{aligned}$$

so η_1 is a strong deformation retraction of D^- onto S^{l-}. On the other hand,

$$\eta_2(u, t) = (1 - t) u + t \pi(u), \quad (u, t) \in S^{l-} \times [0, 1],$$

where

$$\pi : D \backslash \{0\} \to S, \quad u \mapsto \frac{u}{\|u\|_D}$$

is the radial projection onto S, is a strong deformation retraction of S^{l-} onto $\tilde{S}^{l-}$ by the positive homogeneity of ζ and I. Thus,

$$\tilde{H}_{q-1}(D^-) \approx \tilde{H}_{q-1}(\tilde{S}^{l-}). \tag{4.72}$$

Combining (4.69)–(4.72) gives

$$C_q(I, 0) \approx \tilde{H}_{q-1}(\tilde{S}^{l-}). \tag{4.73}$$

If $(a, b) \in I^l$, then $I < 0$ on $S^l \backslash \{0\}$ by the proof of Theorem 4.8.6 (*v*) and hence $\tilde{S}^{l-} = \tilde{S}^l$. Since the latter is homeomorphic to $\tilde{N}_l$, (*ii*) follows. Since I_l and I^{l-1} are subsets of the same connected component of $\mathbb{R}^2 \backslash \Sigma(A)$, (*i*) follows from Proposition 4.11.1 and (*ii*).

Now let $(a, b) \in II_l$. By Theorem 4.8.6 (*iii*), $\tilde{S}^{l-}$ is a proper subset of $\tilde{S}^l$ and hence $C_q(I, 0) \approx \tilde{H}_{q-1}(\tilde{S}^{l-}) = 0$ for $q \geqslant d_l$. Since $\tilde{H}_q(\tilde{S}^l) = \delta_{q(d_l-1)} \mathbb{Z}_2$, the exact sequence

$$\cdots \longrightarrow \tilde{H}_q(\tilde{S}^l) \longrightarrow H_q(\tilde{S}^l, \tilde{S}^{l-}) \longrightarrow \tilde{H}_{q-1}(\tilde{S}^{l-})$$

$$\longrightarrow \tilde{H}_{q-1}(\tilde{S}^l) \longrightarrow \cdots$$

of the pair $(\tilde{S}^l, \tilde{S}^{l-})$ now gives

$$\tilde{H}_{q-1}(\tilde{S}^{l-}) \approx H_q(\tilde{S}^l, \tilde{S}^{l-})/\delta_{q(d_l-1)} \mathbb{Z}_2. \qquad (4.74)$$

By the Poincaré–Lefschetz duality theorem,

$$H_q(\tilde{S}^l, \tilde{S}^{l-}) \approx \check{H}^{d_l-1-q}(\tilde{S}^{l+}). \qquad (4.75)$$

Writing $u \in \tilde{S}^{l+}$ as $\zeta(v + y)$ with $v + y \in N_{l-1} \oplus E_l$, let

$$\eta_3(u, t) = \zeta((1 - t) v + t \eta(y) + y), \quad (u, t) \in \tilde{S}^{l+} \times [0, 1].$$

We have

$I(\eta_3(u, t))$

$\quad = \inf_{w \in M_l} I((1 - t) v + t \eta(y) + y + w) \qquad$ by Proposition 4.7.1 *(ii)*

$\quad \geqslant \inf_{w \in M_l} [(1 - t) I(v + y + w)$

$\qquad\qquad + t I(\eta(y) + y + w)] \qquad$ by Proposition 4.6.1 *(i)*

$\quad \geqslant (1 - t) I(\zeta(v + y)) + t I(\zeta(\eta(y) + y)) \quad$ by Proposition 4.7.1 *(ii)*

$\quad \geqslant I(u) \qquad\qquad\qquad\qquad\qquad$ by Proposition 4.8.1 *(i)*

$\quad \geqslant 0.$

If $\eta_3(u, t) = 0$, then $(1 - t) v + t \eta(y) + y = 0$ and hence $y = 0$, so $u = \zeta(v)$. Since $u \neq 0$, then $v \neq 0$, so

$$I(u) \leqslant I(v) \qquad\qquad \text{by Proposition 4.7.1 } (ii)$$

$$\leqslant \|v\|_D^2 - \underline{\lambda} \|v\|^2 \qquad \text{by (4.19)}$$

$$\leqslant -(\underline{\lambda} - \lambda_{l-1}) \|v\|^2 \quad \text{by (4.5)}$$

$$< 0,$$

contrary to assumption. So $\eta_3(u, t) \neq 0$. Thus,

$$\eta_4 = \pi \circ \eta_3$$

is a strong deformation retraction of $\tilde{S}^{l+}$ onto $\tilde{\mathcal{N}}_l^+$, and hence

$$\check{H}^{d_l-1-q}(\tilde{S}^{l+}) \approx \check{H}^{d_l-1-q}(\tilde{\mathcal{N}}_l^+). \qquad (4.76)$$

Applying the Poincaré–Lefschetz duality theorem again gives

$$\breve{H}^{d_l-1-q}(\tilde{\mathcal{N}}_l^+) \approx H_{q-d_{l-1}}(\tilde{\mathcal{N}}_l, \tilde{\mathcal{N}}_l^-). \tag{4.77}$$

By Theorem 4.8.6 (iii), $\tilde{\mathcal{N}}_l^-$ is a proper subset of $\tilde{\mathcal{N}}_l$ and hence $\tilde{H}_{q-d_{l-1}-1}(\tilde{\mathcal{N}}_l^-)$ $= 0$ for $q \geq d_l$. Since $\tilde{H}_{q-d_{l-1}-1}(\tilde{\mathcal{N}}_l) = \delta_{qd_l} \mathbb{Z}_2$, the exact sequence

$$\cdots \longrightarrow \tilde{H}_{q-d_{l-1}}(\tilde{\mathcal{N}}_l) \longrightarrow H_{q-d_{l-1}}(\tilde{\mathcal{N}}_l, \tilde{\mathcal{N}}_l^-) \longrightarrow \tilde{H}_{q-d_{l-1}-1}(\tilde{\mathcal{N}}_l^-)$$

$$\longrightarrow \tilde{H}_{q-d_{l-1}-1}(\tilde{\mathcal{N}}_l) \longrightarrow \cdots$$

of the pair $(\tilde{\mathcal{N}}_l, \tilde{\mathcal{N}}_l^-)$ then gives

$$H_{q-d_{l-1}}(\tilde{\mathcal{N}}_l, \tilde{\mathcal{N}}_l^-)/\delta_{q(d_l-1)} \mathbb{Z}_2 \approx \tilde{H}_{q-d_{l-1}-1}(\tilde{\mathcal{N}}_l^-). \tag{4.78}$$

Combining (4.73)–(4.78) gives $C_q(I, 0) \approx \tilde{H}_{q-d_{l-1}-1}(\tilde{\mathcal{N}}_l^-)$, from which the rest of (iii) follows. □

Remark 4.11.3 Note that (4.72) holds for all $(a, b) \in Q_l$ and (4.74)–(4.78) hold for all $(a, b) \in \mathrm{II}_l$, so

$$\tilde{H}_{q-1}(D^-) \approx \tilde{H}_{q-d_{l-1}-1}(\tilde{\mathcal{N}}_l^-) \quad \forall (a, b) \in \Sigma(A) \cap \mathrm{II}_l.$$

If $(a, b) \in C_l$, (4.74)–(4.77) hold and $\tilde{\mathcal{N}}_l^- = \varnothing$ by Theorem 4.8.6 (ii), so

$$\tilde{H}_{q-1}(D^-) \approx H_{q-d_{l-1}}(\tilde{\mathcal{N}}_l)/\delta_{q(d_l-1)} \mathbb{Z}_2 = \delta_{qd_{l-1}} \mathbb{Z}_2.$$

If $(a, b) \in C^l$, (4.74)–(4.76) hold and, referring to (4.37), $\tilde{\mathcal{N}}_l^+ = \tilde{K}$ by Theorem 4.8.6 (iv), so

$$\tilde{H}_{q-1}(D^-) \approx \breve{H}^{d_l-1-q}(\tilde{K})/\delta_{q(d_l-1)} \mathbb{Z}_2.$$

Let

$$O_l = \begin{cases} \mathrm{I}_2, & l = 2 \\ \mathrm{I}_l \cup \mathrm{I}^{l-1}, & l \geq 3. \end{cases} \tag{4.79}$$

Then

$$C_q(I, 0) \approx \delta_{qd_{l-1}} \mathbb{Z}_2, \quad (a, b) \in O_l \tag{4.80}$$

by Theorem 4.11.2 (i) and (ii), so the following corollary is now immediate from Proposition 4.11.1 and Theorem 4.11.2 (iii).

Corollary 4.11.4 *The points (a, b) and (a', b') belong to different connected components of $\mathbb{R}^2 \setminus \Sigma(A)$ in the following cases:*

(i) $(a, b) \in O_l$ and $(a', b') \in O_{l'}$ for some $l \neq l'$,

(ii) $(a, b) \in O_l$ and $(a', b') \in \mathrm{II}_{l'}$ for some l, l'.

For example, there is no path in $\mathbb{R}^2 \backslash \Sigma(A)$ joining a point in II_l to the diagonal $a = b$.

Remark 4.11.5 For problem (4.11), Theorem 4.11.2 is due to Dancer [39, 40] and Perera and Schechter [115, 116, 117].

5

Jumping nonlinearities

5.1 Introduction

Consider the equation

$$Au = f(u), \quad u \in D \tag{5.1}$$

where A is as in Chapter 4 and $f \in C(D, H)$ is a potential operator. Solutions of (5.1) coincide with critical points of the C^1-functional

$$G(u) = \|u\|_D^2 - 2F(u), \quad u \in D$$

where F is the potential of f with $F(0) = 0$.

We will first consider the solvability of (5.1) when

$$f(u) = bu^+ - au^- - p(u) \tag{5.2}$$

for some $(a, b) \in \mathbb{R}^2$ and a bounded potential operator $p \in C(D, H)$ with

$$p(u) = o(\|u\|_D) \quad \text{as} \quad \|u\|_D \to \infty. \tag{5.3}$$

We say that (5.1) is resonant at infinity if $(a, b) \in \Sigma(A)$, otherwise it is non-resonant. Now we have

$$G(u) = I(u, a, b) + 2P(u), \quad u \in D \tag{5.4}$$

where P is the potential of p with $P(0) = 0$, and P is bounded and

$$P(u) = \int_0^1 (p(su), u) \, ds = o(\|u\|_D^2) \quad \text{as} \quad \|u\|_D \to \infty \tag{5.5}$$

by Proposition 4.3.2 and (5.1.3).

107

We will also consider the existence of nontrivial solutions when

$$f(u) = b_0 u^+ - a_0 u^- - p_0(u) \tag{5.6}$$

for some $(a_0, b_0) \in \mathbb{R}^2$ and a bounded potential operator $p_0 \in C(D, H)$ with

$$p_0(u) = o(\|u\|_D) \quad \text{as} \quad \|u\|_D \to 0. \tag{5.7}$$

We say that (5.1) is resonant at zero if $(a_0, b_0) \in \Sigma(A)$, otherwise it is nonresonant. We have

$$G(u) = I(u, a_0, b_0) + 2P_0(u), \quad u \in D$$

where P_0 is the potential of p_0 with $P_0(0) = 0$, and P_0 is bounded and

$$P_0(u) = \int_0^1 (p_0(su), u) \, ds = o(\|u\|_D^2) \quad \text{as} \quad \|u\|_D \to 0 \tag{5.8}$$

by Proposition 4.3.2 and (5.1.7).

Example 5.1.1 In problem (4.2.2), 3.1.2 and (5.3) hold if

$$f(x, t) = bt^+ - at^- - p(x, t)$$

for some Carathéodory function p on $\Omega \times \mathbb{R}$ with

$$p(x, t) = o(t) \quad \text{as} \quad |t| \to \infty, \quad \text{uniformly a.e.},$$

and (5.6) and (5.7) also hold when

$$f(x, t) = b_0 t^+ - a_0 t^- - p_0(x, t) \tag{5.9}$$

with

$$p_0(x, t) = o(t) \quad \text{as} \quad t \to 0, \quad \text{uniformly a.e.} \tag{5.10}$$

Here

$$G(u) = \int_\Omega |\nabla u|^2 - a\,(u^-)^2 - b\,(u^+)^2 + 2P(x, u), \quad u \in H_0^1(\Omega)$$

where the primitive

$$P(x, t) = \int_0^t p(x, s) \, ds = o(t^2) \quad \text{as} \quad |t| \to \infty, \quad \text{uniformly a.e.}$$

When (5.9) and (5.10) hold, we also have

$$G(u) = \int_\Omega |\nabla u|^2 - a_0\,(u^-)^2 - b_0\,(u^+)^2 + 2P_0(x, u)$$

where

$$P_0(x, t) = \int_0^t p_0(x, s) \, ds = o(t^2) \quad \text{as} \quad t \to 0, \quad \text{uniformly a.e.} \tag{5.11}$$

5.2 Compactness

In this section we prove some results on the convergence of (PS) and (C) sequences of G. We assume (5.2) and (5.3), so that G is given by (5.4).

Our first lemma implies that every bounded (PS) sequence has a convergent subsequence.

Lemma 5.2.1 *Every bounded sequence* $(u_j) \subset D$ *such that* $G'(u_j) \to 0$ *has a convergent subsequence.*

Proof Since

$$u_j = A^{-1}(bu_j^+ - au_j^- - p(u_j) + G'(u_j)/2) = A^{-1}(u_j')$$

where (u_j') is bounded and A^{-1} is compact, u_j converges in D for a renamed subsequence. $\qquad\square$

The following lemma is useful for verifying the boundedness of (PS) sequences.

Lemma 5.2.2 *If* $G'(u_j) \to 0$ *and* $\rho_j := \|u_j\|_D \to \infty$, *then a subsequence of* $\tilde{u}_j := u_j/\rho_j$ *converges to a nontrivial critical point of* I, *in particular,* $(a, b) \in \Sigma(A)$.

Proof We have

$$I'(\tilde{u}_j) = \frac{I'(u_j)}{\rho_j} = \frac{G'(u_j)}{\rho_j} - 2\frac{p(u_j)}{\|u_j\|_D} \to 0$$

by (5.3) and $\|\tilde{u}_j\|_D = 1$, so the conclusion follows from Lemma 4.4.1. $\qquad\square$

We can now prove the (PS) condition in the nonresonant case.

Proposition 5.2.3 *If* $(a, b) \notin \Sigma(A)$ *and* (5.2) *and* (5.3) *hold, then every sequence* $(u_j) \subset D$ *such that* $G'(u_j) \to 0$ *has a convergent subsequence, in particular,* G *satisfies* (PS).

Proof Since $(a, b) \notin \Sigma(A)$, (u_j) is bounded by Lemma 5.2.2, so the conclusion follows from Lemma 5.2.1. $\qquad\square$

Finally we give sufficient conditions for the (C) condition to hold in the resonant case. The nonquadratic part of G is given by

$$H(u) = G(u) - \frac{1}{2}\left(G'(u), u\right) = 2P(u) - (p(u), u)$$

by (4.38). Note that $(H(u_j))$ is bounded for every (C) sequence (u_j). Denoting by $\mathcal{N}$ the class of sequences $(u_j) \subset D$ such that $\rho_j := \|u_j\|_D \to \infty$ and $\tilde{u}_j := u_j/\rho_j$ converges weakly to some $\tilde{u} \neq 0$, we assume one of

$(H_\pm)$ Every sequence $(u_j) \in \mathcal{N}$ has a subsequence such that $H(u_j) \to \pm\infty$.

In particular, no (C) sequence can belong to $\mathcal{N}$.

Proposition 5.2.4 *If* (5.2), (5.3), *and* (H_+) *or* (H_-) *hold, then G satisfies* (C).

Proof If a (C) sequence (u_j) is unbounded, then Lemma 5.2.2 gives a subsequence that belongs to $\mathcal{N}$, contradicting $(H_\pm)$, so (u_j) is bounded and hence has a convergent subsequence by Lemma 5.2.1. □

Example 5.2.5 In Example 5.1.1,

$$H(u) = \int_\Omega H(x, u)$$

where

$$H(x, t) = 2P(x, t) - tp(x, t),$$

and $(H_\pm)$ holds if

$$H(x, t) \geqslant \text{(resp. } \leqslant \text{)} \, C(x) \quad \text{a.e.} \tag{5.12}$$

for some $C \in L^1(\Omega)$ and

$$H(x, t) \to \pm\infty \quad \text{a.e. as } |t| \to \infty.$$

Indeed, if $(u_j) \in \mathcal{N}$, for a subsequence, $\tilde{u}_j \to \tilde{u}$ a.e. and hence

$$H(u_j) \geqslant \text{(resp. } \leqslant \text{)} \int_{\tilde{u}(x) \neq 0} H(x, \rho_j \tilde{u}_j(x)) + \int_{\tilde{u}(x)=0} C(x) \to \pm\infty$$

by Fatou's lemma.

5.3 Critical groups at infinity

In this section we consider the problem of computing the critical groups of $G = I(\cdot, a, b) + 2P$ at infinity when (5.3) holds.

First we show that in the nonresonant case the lower-order term $2P$ can be deformed away outside a large ball without changing the critical set of G. Let S be the unit sphere in D. Since $(a, b) \notin \Sigma(A)$, $\delta := \inf_{u \in S} \|I'(u)\|_D > 0$ by Lemma 4.4.1, and then $\inf_{u \in S} \|I'(Ru)\|_D = \delta R$ for $R > 0$ by the positive homogeneity of I'. Since $\sup_{u \in S} \|p(Ru)\|_D = o(R)$ by (5.3), it follows that

$$\inf_{u \in S} \|G'(Ru)\|_D = (\delta + o(1)) R \quad \text{as } R \to \infty. \tag{5.13}$$

Take a smooth function $\varphi : [0, \infty) \to [0, 2]$ such that $\varphi = 2$ on $[0, 1]$ and $\varphi = 0$ on $[4, \infty)$ and set

$$\tilde{G}(u) = I(u) + \varphi(\|u\|_D^2 / R^2) \, P(u),$$

so that

$$\tilde{G}(u) = \begin{cases} G(u), & \|u\|_D \leqslant R \\ I(u), & \|u\|_D \geqslant 2R. \end{cases} \tag{5.14}$$

Since $\sup_{u \in S} \left\| \varphi'(\|Ru\|_D^2 / R^2) \, 2Ru/R^2 \right\|_D = O(R^{-1})$ and $\sup_{u \in S} |P(Ru)| = o(R^2)$ by (5.5), (5.13) holds with G replaced by $\tilde{G}$ also. So for sufficiently large R,

$$\inf_{\|u\|_D \geqslant R} \|G'(u)\|_D > 0, \qquad \inf_{\|u\|_D \geqslant R} \|\tilde{G}'(u)\|_D > 0 \tag{5.15}$$

and hence the critical sets of both G and $\tilde{G}$ are in $\overset{\circ}{B}_R$, so solutions of (5.1) coincide with critical points of $\tilde{G}$ by (5.14).

Proposition 5.3.1 *If $(a, b) \notin \Sigma(A)$ and (5.3) holds, then $\tilde{G}$ satisfies* (PS) *and*

$$C_q(\tilde{G}, \infty) \approx C_q(I, 0) \quad \forall q.$$

Proof Every (PS) sequence of $\tilde{G}$ has a subsequence in $\overset{\circ}{B}_R$ by (5.15), which then is a (PS) sequence of G by (5.14) and hence has a convergent subsequence by Proposition 5.2.3.

Since I, P, and φ all map bounded sets into bounded sets, so does $\tilde{G}$. The critical values of $\tilde{G}$ are bounded from below by $\inf \tilde{G}(B_R)$ by (5.15). Taking the a in (1.27) to be less than both $\inf \tilde{G}(B_{2R})$ and $\inf I(B_{2R})$, say a', gives

$$C_q(\tilde{G}, \infty) = H_q(D, \tilde{G}^{a'}) = H_q(D, I^{a'})$$

since $\tilde{G}^{a'}$ and $I^{a'}$ lie outside B_{2R}, where $\tilde{G} = I$ by (5.14). Since the origin is the only critical point of I and $a' < I(0)$,

$$H_q(D, I^{a'}) \approx C_q(I, 0)$$

by Proposition 1.4.1. $\qquad\qquad\qquad\qquad\qquad\qquad\qquad\qquad\qquad\qquad\qquad\qquad\square$

In the resonant case we strengthen the conditions $(H_\pm)$ of Section 5.2 to

$(H_\pm)$ H is bounded from below (resp. above) and every sequence $(u_j) \in \mathcal{N}$ has a subsequence such that

$$H(tu_j) \to \pm\infty \quad \forall t \geqslant 1.$$

Example 5.3.2 In Example 5.2.5, H is bounded from below (resp. above) by (5.12), and if $(u_j) \in \mathcal{N}$, for a subsequence, $\tilde{u}_j \to \tilde{u}$ a.e. and hence

$$H(tu_j) \geq \text{(resp. } \leqslant \text{)} \int_{\tilde{u}(x) \neq 0} H(x, t\rho_j \tilde{u}_j(x)) + \int_{\tilde{u}(x)=0} C(x) \to \pm\infty \quad \forall t \geq 1$$

by Fatou's lemma.

Lemma 5.3.3 *If* $(H_\pm)$ *holds, then* P *is bounded from below (resp. above) and every sequence* $(u_j) \in \mathcal{N}$ *has a subsequence such that*

$$P(u_j) \to \pm\infty. \tag{5.16}$$

Proof We have

$$\frac{d}{dt} \left(-\frac{P(tu)}{t^2} \right) = \frac{H(tu)}{t^3},$$

and

$$\lim_{t \to \infty} \frac{P(tu)}{t^2} = 0$$

by (5.5), so

$$P(u) = \int_1^\infty \frac{H(tu)}{t^3} \, dt.$$

Since $\int_1^\infty dt/t^3 = 1/2$, we have $\inf H/2 \leqslant P \leqslant \sup H/2$, and (5.16) for the subsequence in $(H_\pm)$ follows from Fatou's lemma. □

We can now prove the following.

Proposition 5.3.4 *Let* $(a, b) \in Q_l$ *and assume* (5.3).

(i) *If* $(a, b) \in C_l$ *and* (H_+) *holds, then* $C_{d_{l-1}}(G, \infty) \neq 0$.
(ii) *If* $(a, b) \in C^l$ *and* (H_-) *holds, then* $C_{d_l}(G, \infty) \neq 0$.

Proof (i) Since $(a, b) \in C_l$, $b = \nu_{l-1}(a)$ and hence $n_{l-1}(a, b) = 0$ by Lemma 4.7.8 (i), so $I \geq 0$ on $B = S_l(a, b)$ by Proposition 4.7.5 (i). Since (H_+) holds, P is bounded from below by Lemma 5.3.3, so it follows that G is bounded from below on B. Let $a' < G|_B$ be less than all critical values. For $v \in N_{l-1}$,

$$G(v) \leqslant -(\underline{\lambda}/\lambda_{l-1} - 1 + o(1)) \|v\|_D^2 \quad \text{as} \quad \|v\|_D \to \infty$$

by (4.19), (4.5), and (5.5), and $\underline{\lambda} > \lambda_{l-1}$ since $(a, b) \in Q_l$, so $G \leqslant a'$ on $A = \{v \in N_{l-1} : \|v\|_D = R\}$ for sufficiently large $R > 0$.

By Proposition 2.4.3, A homologically links M_{l-1} in dimension $q = d_{l-1} - 1$. Define a homeomorphism of D by

$$h(u) = v + \sigma(w, a, b), \quad u = v + w \in N_{l-1} \oplus M_{l-1}.$$

Then, noting that $h|_{N_{l-1}} = id_{N_{l-1}}$ since $\sigma(0) = 0$ by positive homogeneity, $h(A) = A$ homologically links $h(M_{l-1}) = \sigma(M_{l-1}) = B$ in dimension q by Proposition 2.4.5.

Since G satisfies (C) by Proposition 5.2.4, the conclusion now follows from Proposition 2.4.7.

(*ii*) For $w \in M_l$,

$$G(w) \geqslant (1 - \bar{\lambda}/\lambda_{l+1} + o(1)) \|w\|_D^2 \quad \text{as} \quad \|w\|_D \to \infty$$

by (4.19), (4.6), and (5.5), and $\bar{\lambda} < \lambda_{l+1}$ since $(a, b) \in Q_l$. Since G maps bounded sets into bounded sets, it follows that G is bounded from below on $B = M_l$. Let $a' < G|_B$ be less than all critical values. We claim that $G \leqslant a'$ on $A = \{\zeta(v, a, b) : v \in N_l, \|v\|_D = R\}$ for sufficiently large $R > 0$. If not, there is a sequence $(v_j) \subset N_l$, $\|v_j\|_D = R_j \to \infty$ such that setting $u_j = \zeta(v_j)$ we have $G(u_j) > a'$. Since N_l is finite dimensional, $\tilde{v}_j := v_j/R_j$ converges to some $\tilde{v} \neq 0$ in N_l for a renamed subsequence. By the continuity of ζ, $\zeta(\tilde{v}_j) \to \zeta(\tilde{v}) = \tilde{v} + \tau(\tilde{v}) \neq 0$. Then

$$\rho_j := \|u_j\|_D = R_j \|\zeta(\tilde{v}_j)\|_D \to \infty$$

and

$$\tilde{u}_j := \frac{u_j}{\rho_j} = \frac{\zeta(\tilde{v}_j)}{\|\zeta(\tilde{v}_j)\|_D} \to \frac{\zeta(\tilde{v})}{\|\zeta(\tilde{v})\|_D} \neq 0,$$

so $(u_j) \in \mathcal{N}$. Since (H_-) holds, then $P(u_j) \to -\infty$ for a renamed subsequence by Lemma 5.3.3. Since $(a, b) \in C^l$, $b = \mu_l(a)$ and hence $m_l(a, b) = 0$ by Lemma 4.7.8 (*ii*), so $I(u_j) \leqslant 0$ by Proposition 4.7.5 (*ii*) and it follows that $G(u_j) \to -\infty$, a contradiction.

By Proposition 2.4.3, $\{v \in N_l : \|v\|_D = R\}$ homologically links B in dimension $q = d_l - 1$. Define a homeomorphism of D by

$$h(u) = \zeta(v, a, b) + w, \quad u = v + w \in N_l \oplus M_l.$$

Then, noting that $h|_{M_l} = id_{M_l}$ since $\zeta(0) = 0$ by positive homogeneity, $h(\{v \in N_l : \|v\|_D = R\}) = A$ homologically links $h(B) = B$ in dimension q by Proposition 2.4.5.

Since G satisfies (C) by Proposition 5.2.4, the conclusion now follows from Proposition 2.4.7. $\qquad \square$

Critical groups can be computed more precisely when (H_+) holds. Referring to (4.37) and (4.68), first we prove a lemma.

Lemma 5.3.5 *If $(a, b) \in Q_l$ and (5.3) and (H_+) hold, then*

$$C_q(G, \infty) \approx \tilde{H}_{q-1}(D^-) \quad \forall q.$$

Proof For $u \in S$ and $t > 0$,

$$G(tu) = t^2 I(u) + 2P(tu) \tag{5.17}$$

since I is positive homogeneous of degree 2, so

$$\frac{d}{dt}\left(G(tu)\right) = 2\left(t\, I(u) + (p(tu), u)\right)$$

$$= \frac{2}{t}\left(G(tu) - H(tu)\right)$$

$$\leqslant \frac{2}{t}\left(G(tu) - \inf H\right) \tag{5.18}$$

and hence all critical values of G are greater than or equal to inf H (note that inf $H \leqslant 0$ since $H(0) = 0$). So

$$C_q(G, \infty) \approx \tilde{H}_{q-1}(G^a)$$

for any $a < \inf H$ by Proposition 1.4.5 *(ii)*. On D^+, $G = I + 2P \geqslant \inf H$ by the proof of Lemma 5.3.3, so $G^a \subset D^-$. We will show that G^a is a strong deformation retract of D^-.

For $u \in S^-$ and $t > 0$,

$$G(tu) = t^2\left(I(u) + \frac{2P(tu)}{t^2}\right) \to -\infty \quad \text{as} \quad t \to \infty$$

by (5.17) and (5.5), so $G(tu) \leqslant a$ for all sufficiently large t. By (5.18),

$$G(tu) \leqslant a \implies \frac{d}{dt}\left(G(tu)\right) < 0.$$

Thus, there is a unique $T_a(u) > 0$ such that

$$t < (\text{resp.} =, >) \; T_a(u) \implies G(tu) > (\text{resp.} =, <) \; a$$

and the map $T_a : S^- \to (0, \infty)$ is C^1 by the implicit function theorem. Then

$$G^a = \{tu : u \in S^-, \; t \geqslant T_a(u)\}$$

and

$$D^- \times [0, 1] \to D^-, \quad (u, t) \mapsto \begin{cases} (1 - t)u + t\, T_a(\pi(u))\, \pi(u), & u \in D^- \backslash G^a \\ u, & u \in G^a, \end{cases}$$

where π is the radial projection onto S, is a strong deformation retraction of D^- onto G^a. $\qquad\qquad\qquad\qquad\qquad\qquad\qquad\qquad\qquad\qquad\qquad$ □

We can now prove the following.

Proposition 5.3.6 *Let* $(a, b) \in Q_l \cap \Sigma(A)$ *and assume* (5.3) *and* (H_+).

(i) *If* $(a, b) \in C_l$, *then*

$$C_q(G, \infty) \approx \delta_{qd_{l-1}} \mathbb{Z}_2.$$

(ii) *If* $(a, b) \in C^l$, *then*

$$C_q(G, \infty) \approx \check{H}^{d_l - 1 - q}(\tilde{K})/\delta_{q(d_l - 1)} \mathbb{Z}_2 \quad \forall q.$$

In particular,

$$C_q(G, \infty) = 0, \quad q < d_{l-1} \text{ or } q \geqslant d_l$$

and $C_{d_{l-1}}(G, \infty) = 0$ *when* $(a, b) \notin C_l$.

(iii) *If* $(a, b) \in \Pi_l$, *then*

$$C_q(G, \infty) \approx \tilde{H}_{q-d_{l-1}-1}(\tilde{\mathcal{N}}_l^-) \quad \forall q.$$

In particular,

$$C_q(G, \infty) = 0, \quad q \leqslant d_{l-1} \text{ or } q \geqslant d_l$$

and $C_q(G, \infty) = 0$ *for all* q *when* λ_l *is simple.*

Proof Part (i) and the first parts of (ii) and (iii) are immediate from Lemma 5.3.5 and Remark 4.11.3. The second part of (ii) follows since $\tilde{K}$ is a subset of the $(d_l - d_{l-1} - 1)$-dimensional sphere $\tilde{\mathcal{N}}_l$, and a proper subset when $(a, b) \notin C_l$ by Theorem 4.8.7. The second part of (iii) also follows since $\tilde{\mathcal{N}}_l^-$ is a nonempty proper subset of $\tilde{\mathcal{N}}_l$ by Theorem 4.8.6 (iii). $\qquad\qquad\qquad\qquad$ □

5.4 Solvability

In this section we consider the solvability of equation (5.1) when (5.2) and (5.3) hold.

First we consider the nonresonant case. Referring to (4.79), let $\tilde{O}_l$ be the connected component of $\mathbb{R}^2 \backslash \Sigma(A)$ containing O_l.

Theorem 5.4.1 *If* $(a, b) \in \tilde{O}_l$ *for some* l *and* (5.2) *and* (5.3) *hold, then* (5.1) *has a solution. In particular,* (5.1) *has a solution if there is a path in* $\mathbb{R}^2 \backslash \Sigma(A)$ *joining* (a, b) *to the diagonal.*

Proof Let $\widetilde{G}$ be the functional constructed in Section 5.3. By Propositions 5.3.1 and 4.11.1, and (4.80), $\widetilde{G}$ satisfies (PS) and

$$C_q(\widetilde{G}, \infty) \approx C_q(I, 0) \approx \delta_{qd_{l-1}} \mathbb{Z}_2.$$

Since $C_{d_{l-1}}(\widetilde{G}, \infty) \neq 0$, then $\widetilde{G}$ has a critical point by Proposition 1.4.7. $\qquad\square$

In the resonant case we assume the conditions $(H_\pm)$ of Section 5.3.

Theorem 5.4.2 *If $(a, b) \in Q_l$ and (5.2) and (5.3) hold, then (5.1) has a solution in the following cases:*

(i) $(a, b) \in C_l$ and (H_+) holds,
(ii) $(a, b) \in C^l$ and (H_-) holds.

Proof By Propositions 5.2.4 and 5.3.4, G satisfies (C) and $C_{d_{l-1}}(G, \infty) \neq 0$ in Case (i) and $C_{d_l}(G, \infty) \neq 0$ in Case (ii), so G has a critical point by Proposition 1.4.7. $\qquad\square$

5.5 Critical groups at zero

In this section we consider the problem of computing the critical groups of $G = I(\cdot, a_0, b_0) + 2P_0$ at zero when (5.7) holds.

First we show that in the nonresonant case zero is an isolated critical point and the higher-order term $2P_0$ can be deformed away without changing the critical groups there.

Proposition 5.5.1 *If $(a_0, b_0) \notin \Sigma(A)$ and (5.7) holds, then*

$$C_q(G, 0) \approx C_q(I(\cdot, a_0, b_0), 0) \quad \forall q.$$

Proof We apply Theorem 1.4.4 with

$$G_t(u) = I(u, a_0, b_0) + 2(1 - t) P_0(u), \quad u \in D, \ t \in [0, 1],$$

which satisfies (PS) by Proposition 5.2.3 since $(a_0, b_0) \notin \Sigma(A)$. We claim that (ii) holds for a sufficiently small closed and bounded neighborhood U of the origin. If not, there are sequences $(t_j) \subset [0, 1]$ and $(u_j) \subset D \backslash \{0\}$, $\rho_j := \|u_j\|_D \to 0$ such that $G'_{t_j}(u_j) = 0$. Then setting $\widetilde{u}_j := u_j / \rho_j$ we have

$$I'(\widetilde{u}_j, a_0, b_0) = \frac{I'(u_j, a_0, b_0)}{\rho_j} = \frac{G'_{t_j}(u_j)}{\rho_j} - 2(1 - t_j) \frac{p_0(u_j)}{\|u_j\|_D} \to 0$$

by (5.7) and $\|\widetilde{u}_j\|_D = 1$, so a subsequence of $(\widetilde{u}_j)$ converges to a nontrivial critical point of $I(\cdot, a_0, b_0)$ by Lemma 4.4.1, contradicting $(a_0, b_0) \notin \Sigma(A)$.

For $t, t_0 \in [0, 1]$, $u \in U$,

$$|G_t(u) - G_{t_0}(u)| + \|G'_t(u) - G'_{t_0}(u)\|_D$$

$$= 2|t - t_0| (|P_0(u)| + \|p_0(u)\|_D)$$

$$\leqslant C|t - t_0|$$

for some constant $C > 0$ since both P_0 and p_0 are bounded on U, so (*ii*) also holds. □

In the resonant case we assume that zero is an isolated critical point and use a generalized local linking to obtain a nontrivial critical group.

Proposition 5.5.2 *Let* $(a_0, b_0) \in Q_l$ *and assume* (5.8).

(*i*) *If* $(a_0, b_0) \in C_l$ *and there is an* $r > 0$ *such that*

$$P_0(\sigma(w, a_0, b_0)) > 0 \quad \forall w \in M_{l-1}, \ 0 < \|w\|_D \leqslant r, \tag{5.19}$$

   then $C_{d_{l-1}}(G, 0) \neq 0$.

(*ii*) *If* $(a_0, b_0) \in C^l$ *and there is an* $r > 0$ *such that*

$$P_0(\zeta(v, a_0, b_0)) \leqslant 0 \quad \forall v \in N_l, \ \|v\|_D \leqslant r, \tag{5.20}$$

   then $C_{d_l}(G, 0) \neq 0$.

Proof (*i*) We have the direct sum decomposition $E = N_{l-1} \oplus M_{l-1}$, $u = v + w$. Define an origin preserving homeomorphism h from $C = \{u \in E : \|v\|_D \leqslant r, \|w\|_D \leqslant r\}$ onto a neighborhood U of zero by

$$h(u) = v + \sigma(w, a_0, b_0).$$

Since $(a_0, b_0) \in C_l$, $b_0 = v_{l-1}(a_0)$ and hence $n_{l-1}(a_0, b_0) = 0$ by Lemma 4.7.8 (*i*), so $I(\sigma(\cdot), a_0, b_0) \geqslant 0$ on M_{l-1} by Proposition 4.7.5 (*i*). This together with (5.19) gives $G|_{h(C \cap M_{l-1}) \setminus \{0\}} > 0$. By (4.19), (4.5), and (5.8), for $v \in N_{l-1}$,

$$G(v) \leqslant -(\underline{\lambda}/\lambda_{l-1} - 1 + o(1)) \|v\|_D^2 \quad \text{as} \quad \|v\|_D \to 0$$

where $\underline{\lambda} = \min\{a_0, b_0\} > \lambda_{l-1}$ since $(a_0, b_0) \in Q_l$, so $G \leqslant 0$ on $C \cap N_{l-1}$ if r is sufficiently small. Since $\sigma(0) = 0$ by positive homogeneity, $h|_{C \cap N_{l-1}} = id_{C \cap N_{l-1}}$, so this gives $G|_{h(C \cap N_{l-1})} \leqslant 0$. Thus, G has a generalized local linking near zero in dimension $q = d_{l-1}$, and then Proposition 1.9.2 gives the conclusion.

(*ii*) We have the direct sum decomposition $E = N_l \oplus M_l$, $u = v + w$. Define an origin preserving homeomorphism h from $C = \{u \in E : \|v\|_D \leqslant$

$r, \|w\|_D \leqslant r\}$ onto a neighborhood U of zero by

$$h(u) = \zeta(v, a_0, b_0) + w.$$

Since $(a_0, b_0) \in C^l$, $b_0 = \mu_l(a_0)$ and hence $m_l(a_0, b_0) = 0$ by Lemma 4.7.8 (*ii*), so $I(\zeta(\cdot), a_0, b_0) \leqslant 0$ on N_l by Proposition 4.7.5 (*ii*). This together with (5.20) gives $G|_{h(C \cap N_l)} \leqslant 0$. By (4.19), (4.6), and (5.8), for $w \in M_l$,

$$G(w) \geqslant (1 - \overline{\lambda}/\lambda_{l+1} + o(1)) \|w\|_D^2 \quad \text{as} \quad \|w\|_D \to 0$$

where $\overline{\lambda} = \max\{a_0, b_0\} < \lambda_{l+1}$ since $(a_0, b_0) \in Q_l$, so $G > 0$ on $C \cap M_l \setminus \{0\}$ if r is sufficiently small. Since $\zeta(0) = 0$ by positive homogeneity, $h|_{C \cap M_l} = id_{C \cap M_l}$, so this gives $G|_{h(C \cap M_l) \setminus \{0\}} > 0$. Thus, G has a generalized local linking near zero in dimension $q = d_l$, and then Proposition 1.9.2 gives the conclusion. $\qquad \square$

Example 5.5.3 In Example 5.1.1, (5.11) implies (5.8). Clearly, (5.19) holds if $P_0 > 0$ on $\Omega \times (\mathbb{R} \setminus \{0\})$ and (5.20) holds if $P_0 \leqslant 0$ on $\Omega \times \mathbb{R}$. However, local sign conditions are sufficient to obtain the conclusions of Proposition 5.5.2. If there is a $\delta > 0$ such that

$$P_0(x, t) > 0 \quad \forall x \in \Omega, \ 0 < |t| \leqslant \delta, \tag{5.21}$$

then $C_{d_{l-1}}(G, 0) \neq 0$, and if there is a $\delta > 0$ such that

$$P_0(x, t) \leqslant 0 \quad \forall x \in \Omega, \ |t| \leqslant \delta, \tag{5.22}$$

then $C_{d_l}(G, 0) \neq 0$. To see this, take a smooth nondecreasing function $\vartheta : \mathbb{R} \to [-\delta, \delta]$ such that $\vartheta(t) = -\delta$ for $t \leqslant -\delta$, $\vartheta(t) = t$ for $-\delta/2 \leqslant t \leqslant \delta/2$, and $\vartheta(t) = \delta$ for $t \geqslant \delta$, set $\widetilde{P}_0(x, t) = P_0(x, \vartheta(t))$, and apply Theorem 3.2.2 to G and

$$\widetilde{G}(u) = \int_\Omega |\nabla u|^2 - a_0 (u^-)^2 - b_0 (u^+)^2 + 2\widetilde{P}_0(x, u)$$

to get $C_*(G, 0) \approx C_*(\widetilde{G}, 0)$. Since (5.21) implies $\widetilde{P}_0 > 0$ on $\Omega \times (\mathbb{R} \setminus \{0\})$ and (5.22) implies $\widetilde{P}_0 \leqslant 0$ on $\Omega \times \mathbb{R}$, the conclusions follow.

Note that

$$H(u) = 2P_0(u) - (p_0(u), u). \tag{5.23}$$

Critical groups can be computed more precisely when

$$H(u) < 0 \quad \forall u \in B_r \setminus \{0\} \tag{5.24}$$

for some $r > 0$. First we prove a lemma.

Lemma 5.5.4 *If* (5.24) *holds, then*

$$P_0(u) > 0 \quad \forall u \in B_r \backslash \{0\}.$$

Proof We have

$$\frac{d}{dt}\left(\frac{P_0(tu)}{t^2}\right) = -\frac{H(tu)}{t^3}$$

by (5.23), and

$$\lim_{t \to 0} \frac{P_0(tu)}{t^2} = 0$$

by (5.8), so

$$P_0(u) = -\int_0^1 \frac{H(tu)}{t^3} \, dt > 0 \quad \forall u \in B_r \backslash \{0\}$$

by (5.24). $\qquad\qquad\qquad\qquad\qquad\qquad\qquad\qquad\qquad\qquad\qquad\square$

Setting

$$B_- = \{u \in B : I(u, a_0, b_0) < 0\}, \qquad B_+ = B \backslash B_-$$

for $B \subset D$, next we prove a lemma.

Lemma 5.5.5 *If* $(a_0, b_0) \in Q_l$ *and* (5.8) *and* (5.24) *hold, then*

$$C_q(G, 0) \approx \tilde{H}_{q-1}(D_-) \quad \forall q.$$

Proof We have

$$C_q(G, 0) = H_q(G^0 \cap B_r, G^0 \cap B_r \backslash \{0\}).$$

On $B_r \cap D_+ \backslash \{0\}$, $G = I(\cdot, a_0, b_0) + 2P_0 > 0$ by Lemma 5.5.4, so $G^0 \cap B_r \backslash \{0\} \subset D_-$. We will show that $G^0 \cap B_r$ is contractible to 0 and $G^0 \cap B_r \backslash \{0\}$ is a strong deformation retract of D_-. The conclusion will then follow from Lemma 1.4.6 (*ii*).

For $u \in S_r = \partial B_r$ and $0 < t \leqslant 1$,

$$G(tu) = t^2 I(u, a_0, b_0) + 2P_0(tu) \tag{5.25}$$

since I is positive homogeneous of degree 2, so

$$\frac{d}{dt}\left(G(tu)\right) = 2\left(t\, I(u, a_0, b_0) + (p_0(tu), u)\right)$$

$$= \frac{2}{t}\left(G(tu) - H(tu)\right)$$

$$> \frac{2}{t} G(tu) \tag{5.26}$$

by (5.24). So for $u \in (S_r)_-$,

$$G(tu) = t^2 \left(I(u, a_0, b_0) + \frac{2P_0(tu)}{t^2} \right) < 0$$

for all sufficiently small t by (5.25) and (5.8),

$$G(tu) \geqslant 0 \implies \frac{d}{dt} \left(G(tu) \right) > 0$$

by (5.26), and hence there is a unique $0 < T(u) \leqslant 1$ such that $G(tu) < 0$ for $0 < t < T(u)$, $G(T(u) u) \leqslant 0$, and $G(tu) > 0$ for $T(u) < t \leqslant 1$.

We claim that the map $T : (S_r)_- \to (0, 1]$ is continuous. By (5.26) and the implicit function theorem, T is C^1 on $\{u \in (S_r)_- : T(u) < 1\}$, so it suffices to show that if $u_j \to u$ and $T(u) = 1$, then $T(u_j) \to 1$. But for any $t < 1$, $G(tu_j) \to G(tu) < 0$, so $T(u_j) > t$ for j sufficiently large.

Thus,

$$G^0 \cap B_r = \left\{ tu : u \in (S_r)_-, \, 0 \leqslant t \leqslant T(u) \right\}$$

and is radially contractible to 0, and

$$D_- \times [0, 1] \to D_-,$$

$$(u, t) \mapsto \begin{cases} (1 - t) u + t \, T(\pi_r(u)) \, \pi_r(u), & u \in D_- \backslash (G^0 \cap B_r \backslash \{0\}) \\ u, & u \in G^0 \cap B_r \backslash \{0\}, \end{cases}$$

where π_r is the radial projection onto S_r, is a strong deformation retraction of D_- onto $G^0 \cap B_r \backslash \{0\}$. $\qquad \square$

Referring to (4.37), we can now prove

Proposition 5.5.6 *Let* $(a_0, b_0) \in Q_l \cap \Sigma(A)$ *and assume* (5.8) *and* (5.24).

(i) *If* $(a_0, b_0) \in C_l$, *then*

$$C_q(G, 0) \approx \delta_{qd_{l-1}} \mathbb{Z}_2.$$

(ii) *If* $(a_0, b_0) \in C^l$, *then*

$$C_q(G, 0) \approx \check{H}^{d_l - 1 - q} (\tilde{K}(a_0, b_0)) / \delta_{q(d_l - 1)} \mathbb{Z}_2 \quad \forall q.$$

In particular,

$$C_q(G, 0) = 0, \quad q < d_{l-1} \text{ or } q \geqslant d_l$$

and $C_{d_{l-1}}(G, 0) = 0$ *when* $(a_0, b_0) \notin C_l$.

(*iii*) *If* $(a_0, b_0) \in \mathrm{II}_l$, *then*

$$C_q(G, 0) \approx \tilde{H}_{q-d_{l-1}-1}(\tilde{\mathcal{N}}_l(a_0, b_0)_-) \quad \forall q.$$

In particular,

$$C_q(G, 0) = 0, \quad q \leqslant d_{l-1} \text{ or } q \geqslant d_l$$

and $C_q(G, 0) = 0$ *for all* q *when* λ_l *is simple.*

Proof Part (*i*) and the first parts of (*ii*) and (*iii*) are immediate from Lemma 5.5.5 and Remark 4.11.3. The second part of (*ii*) follows since $\tilde{K}(a_0, b_0)$ is a subset of the $(d_l - d_{l-1} - 1)$-dimensional sphere $\tilde{\mathcal{N}}_l(a_0, b_0)$, and a proper subset when $(a_0, b_0) \notin C_l$ by Theorem 4.8.7. The second part of (*iii*) also follows since $\tilde{\mathcal{N}}_l(a_0, b_0)_-$ is a nonempty proper subset of $\tilde{\mathcal{N}}_l(a_0, b_0)$ by Theorem 4.8.6 (*iii*). □

Example 5.5.7 In Example 5.1.1, (5.24) holds if $H < 0$ on $\Omega \times (\mathbb{R} \backslash \{0\})$, but the conclusions of Proposition 5.5.6 also hold under (5.11) and the local sign condition

$$H(x, t) < 0 \quad \forall x \in \Omega, \ 0 < |t| \leqslant \delta \tag{5.27}$$

for some $\delta > 0$. To see this, set

$$\tilde{p}_0(x, t) = \begin{cases} -p_0(x, -\delta)\dfrac{t}{\delta}, & t < -\delta \\ p_0(x, t), & |t| \leqslant \delta \\ p_0(x, \delta)\dfrac{t}{\delta}, & t > \delta, \end{cases} \qquad \tilde{P}_0(x, t) = \int_0^t \tilde{p}_0(x, s)\, ds$$

and apply Theorem 3.2.2 to G and

$$\tilde{G}(u) = \int_\Omega |\nabla u|^2 - a_0 (u^-)^2 - b_0 (u^+)^2 + 2\tilde{P}_0(x, u)$$

to get $C_*(G, 0) \approx C_*(\tilde{G}, 0)$. A simple calculation shows that

$$\tilde{H}(x, t) = 2\tilde{P}_0(x, t) - t\, \tilde{p}_0(x, t) = \begin{cases} H(x, -\delta), & t < -\delta \\ H(x, t), & |t| \leqslant \delta \\ H(x, \delta), & t > \delta, \end{cases}$$

so (5.27) implies $\tilde{H} < 0$ on $\Omega \times (\mathbb{R} \backslash \{0\})$.

5.6 Nonlinearities crossing the Fučík spectrum

In this section we consider the existence of nontrivial solutions of the equation (5.1) when (5.2), (5.3), (5.6), and (5.7) or (5.8) hold.

First we consider the nonresonant case. Let $\widetilde{O}_l$ be as in Section 5.4 and let $\widetilde{\mathrm{II}}_l$ be the set of points that can be joined to II_l by a path in $\mathbb{R}^2 \backslash \Sigma(A)$.

Theorem 5.6.1 *If* (5.2), (5.3), (5.6), *and* (5.7) *hold, then* (5.1) *has a nontrivial solution in the following cases:*

(i) $(a_0, b_0) \in \widetilde{O}_l$ *and* $(a, b) \in \widetilde{O}_{l'}$ *for some* $l \neq l'$,
(ii) $(a_0, b_0) \in \widetilde{O}_l$ *and* $(a, b) \in \widetilde{\mathrm{II}}_{l'}$ *for some* l, l',
(iii) $(a_0, b_0) \in \widetilde{\mathrm{II}}_l$ *and* $(a, b) \in \widetilde{O}_{l'}$ *for some* l, l'.

Proof Let $\widetilde{G}$ be the functional constructed in Section 5.3. By (5.14) and Proposition 5.5.1,

$$C_q(\widetilde{G}, 0) = C_q(G, 0) \approx C_q(I(\cdot, a_0, b_0), 0).$$

By Proposition 5.3.1,

$$C_q(\widetilde{G}, \infty) \approx C_q(I(\cdot, a, b), 0)$$

and $\widetilde{G}$ satisfies (PS). So $\widetilde{G}$ has a nontrivial critical point if $C_q(I(\cdot, a_0, b_0), 0) \not\approx C_q(I(\cdot, a, b), 0)$ for some q by Proposition 1.6.1.

(i) $C_q(I(\cdot, a_0, b_0), 0) \approx \delta_{q d_{l-1}} \mathbb{Z}_2$ and $C_q(I(\cdot, a, b), 0) \approx \delta_{q d_{l'-1}} \mathbb{Z}_2$ by Proposition 4.11.1 and (4.80).

(ii) $C_q(I(\cdot, a_0, b_0), 0) \approx \delta_{q d_{l-1}} \mathbb{Z}_2$ and $C_{d_{l-1}}(I(\cdot, a, b), 0) = 0$ by Proposition 4.11.1 and Theorem 4.11.2 (iii).

(iii) $C_{d_{l'-1}}(I(\cdot, a_0, b_0), 0) = 0$ and $C_q(I(\cdot, a, b), 0) \approx \delta_{q d_{l'-1}} \mathbb{Z}_2$. ☐

Remark 5.6.2 In particular, there is a nontrivial solution when (a_0, b_0) and (a, b) are on opposite sides of C_l or C^l in $Q_l \backslash \Sigma(A)$. For problem (4.11), this was proved by Perera and Schechter [116]. It generalizes a well-known result of Amann and Zehnder [3] on the existence of nontrivial solutions for problems crossing an eigenvalue.

In the case of resonance at zero we have the following theorem.

Theorem 5.6.3 *If* (5.2), (5.3), (5.6), *and* (5.8) *hold, then* (5.1) *has a nontrivial solution in the following cases:*

(i) $(a_0, b_0) \in C_l$, *there is an* $r > 0$ *such that*

$$P_0(\sigma(w, a_0, b_0)) > 0 \quad \forall w \in M_{l-1}, \ 0 < \|w\|_D \leqslant r,$$

and either $(a, b) \in \widetilde{O}_{l'}$ *for some* $l' \neq l$ *or* $(a, b) \in \widetilde{\mathrm{II}}_{l'}$ *for some* l',

(ii) $(a_0, b_0) \in Q_l \cap \Sigma(A) \backslash C_l$, there is an $r > 0$ such that

$$H(u) < 0 \quad \forall u \in D, \ 0 < \|u\|_D \leqslant r,$$

and $(a, b) \in \tilde{O}_{l'}$ for some l',
(iii) $(a_0, b_0) \in C^l$, there is an $r > 0$ such that

$$P_0(\zeta(v, a_0, b_0)) \leqslant 0 \quad \forall v \in N_l, \ \|v\|_D \leqslant r,$$

and either $(a, b) \in \tilde{O}_{l'}$ for some $l' \neq l + 1$ or $(a, b) \in \tilde{\Pi}_{l'}$ for some l'.

Proof Let $\tilde{G}$ be the functional constructed in Section 5.3. By (5.14),

$$C_q(\tilde{G}, 0) = C_q(G, 0).$$

By Proposition 5.3.1,

$$C_q(\tilde{G}, \infty) \approx C_q(I(\cdot, a, b), 0)$$

and $\tilde{G}$ satisfies (PS). So $\tilde{G}$ has a nontrivial critical point if $C_q(G, 0) \not\approx$ $C_q(I(\cdot, a, b), 0)$ for some q by Proposition 1.6.1.

(i) $C_{d_{l-1}}(G, 0) \neq 0$ by Proposition 5.5.2 (i) and $C_{d_{l-1}}(I(\cdot, a, b), 0) = 0$ by Proposition 4.11.1, (4.80), and Theorem 4.11.2 (iii).

(ii) $C_{d_{l'-1}}(G, 0) = 0$ by Proposition 5.5.6 (ii) and (iii), and $C_q(I(\cdot, a, b), 0) \approx \delta_{qd_{l'-1}} \mathbb{Z}_2$.

(iii) $C_{d_l}(G, 0) \neq 0$ by Proposition 5.5.2 (ii) and $C_{d_l}(I(\cdot, a, b), 0) = 0$. $\quad\square$

In the case of resonance at infinity we assume the conditions $(H_\pm)$ of Section 5.3.

Theorem 5.6.4 *If (5.2), (5.3), (5.6), and (5.7) hold, then (5.1) has a nontrivial solution in the following cases:*

(i) $(a, b) \in C_l$, (H_+) holds, and either $(a_0, b_0) \in \tilde{O}_{l'}$ for some $l' \neq l$ or $(a_0, b_0) \in \tilde{\Pi}_{l'}$ for some l',
(ii) $(a, b) \in Q_l \cap \Sigma(A) \backslash C_l$, (H_+) holds, and $(a_0, b_0) \in \tilde{O}_{l'}$ for some l',
(iii) $(a, b) \in C^l$, (H_-) holds, and either $(a_0, b_0) \in \tilde{O}_{l'}$ for some $l' \neq l + 1$ or $(a_0, b_0) \in \tilde{\Pi}_{l'}$ for some l'.

Proof By Proposition 5.5.1,

$$C_q(G, 0) \approx C_q(I(\cdot, a_0, b_0), 0).$$

By Proposition 5.2.4, G satisfies (C). So G has a nontrivial critical point if $C_q(G, \infty) \not\approx C_q(I(\cdot, a_0, b_0), 0)$ for some q by Proposition 1.6.1.

(i) $C_{d_{l-1}}(G, \infty) \neq 0$ by Proposition 5.3.4 (i) and $C_{d_{l-1}}(I(\cdot, a_0, b_0), 0) = 0$ by Proposition 4.11.1, (4.80), and Theorem 4.11.2 (iii).

(ii) $C_{d_{l'-1}}(G, \infty) = 0$ by Proposition 5.3.6 (ii) and (iii), and $C_q(I(\cdot, a_0, b_0), 0) \approx \delta_{qd_{l'-1}} \mathbb{Z}_2$.

(iii) $C_{d_l}(G, \infty) \neq 0$ by Proposition 5.3.4 (ii) and $C_{d_l}(I(\cdot, a_0, b_0), 0) = 0$. $\quad\square$

In the double resonance case we have the following.

Theorem 5.6.5 *If* (5.2), (5.3), (H_+), (5.6), *and* (5.8) *hold, then* (5.1) *has a nontrivial solution in the following cases:*

(i) $(a_0, b_0) \in C_l$, *there is an* $r > 0$ *such that*

$$P_0(\sigma(w, a_0, b_0)) > 0 \quad \forall w \in M_{l-1}, 0 < \|w\|_D \leqslant r,$$

and either $(a, b) \in C_{l'}$ *for some* $l' \neq l$ *or* $(a, b) \in Q_{l'} \cap \Sigma(A)\backslash C_{l'}$ *for some* l',

(ii) $(a_0, b_0) \in C^l$, *there is an* $r > 0$ *such that*

$$P_0(\zeta(v, a_0, b_0)) \leqslant 0 \quad \forall v \in N_l, \|v\|_D \leqslant r,$$

and either $(a, b) \in C_{l'}$ *for some* $l' \neq l + 1$ *or* $(a, b) \in Q_{l'} \cap \Sigma(A)\backslash C_{l'}$ *for some* l'.

Proof By Proposition 5.2.4, G satisfies (C). We apply Proposition 1.6.1.

(i) $C_{d_{l-1}}(G, 0) \neq 0$ by Proposition 5.5.2 (i) and $C_{d_{l-1}}(G, \infty) = 0$ by Proposition 5.3.6.

(ii) $C_{d_l}(G, 0) \neq 0$ by Proposition 5.5.2 (ii) and $C_{d_l}(G, \infty) = 0$. $\quad\square$

Theorem 5.6.6 *If* (5.2), (5.3), (5.6), *and* (5.8) *hold and there is an* $r > 0$ *such that*

$$H(u) < 0 \quad \forall u \in D, 0 < \|u\|_D \leqslant r,$$

then (5.1) *has a nontrivial solution in the following cases:*

(i) $(a, b) \in C_l$, (H_+) *holds, and either* $(a_0, b_0) \in C_{l'}$ *for some* $l' \neq l$ *or* $(a_0, b_0) \in Q_{l'} \cap \Sigma(A)\backslash C_{l'}$ *for some* l',

(ii) $(a, b) \in C^l$, (H_-) *holds, and either* $(a_0, b_0) \in C_{l'}$ *for some* $l' \neq l + 1$ *or* $(a_0, b_0) \in Q_{l'} \cap \Sigma(A)\backslash C_{l'}$ *for some* l'.

Proof By Proposition 5.2.4, G satisfies (C). We apply Proposition 1.6.1.

(i) $C_{d_{l-1}}(G, \infty) \neq 0$ by Proposition 5.3.4 (i) and $C_{d_{l-1}}(G, 0) = 0$ by Proposition 5.5.6.

(ii) $C_{d_l}(G, \infty) \neq 0$ by Proposition 5.3.4 (ii) and $C_{d_l}(G, 0) = 0$. $\quad\square$

6

Sandwich pairs

6.1 Introduction

The notion of sandwich pairs is a useful tool for finding critical points of a functional. It was introduced by Schechter [145, 146] and was based on the sandwich theorem for complementing subspaces by Schechter [137, 138] and Silva [150].

Definition 6.1.1 We say that a pair of subsets A, B of a Banach space E is a sandwich pair if every $G \in C^1(E, \mathbb{R})$ satisfying

$$-\infty < b := \inf_B G \leqslant \sup_A G =: a < +\infty \qquad (6.1)$$

and $(PS)_c$ for all $c \in [b, a]$ has a critical point u with $b \leqslant G(u) \leqslant a$.

Example 6.1.2 If $E = N \oplus M$ is a direct sum decomposition with N nontrivial and finite dimensional, then N, M form a sandwich pair. In fact, we will see in Section 6.3 that this is a special case of a much more general class of sandwich pairs.

6.2 Flows

First we give a criterion, involving a certain class of flows on E, for a pair of subsets to form a sandwich pair.

Denote by Σ the set of all maps $\sigma \in C(E \times [0, 1], E)$ such that, writing $\sigma(u, t) = \sigma(t) u$,

(i) $\sigma(0) = id_E$,

(ii) $\displaystyle \sup_{(u,t) \in E \times [0,1]} \| \sigma(t) u - u \| < \infty.$

125

Theorem 6.2.1 *A, $B \subset E$ form a sandwich pair if*

$$\sigma(1) A \cap B \neq \varnothing \quad \forall \sigma \in \Sigma. \tag{6.2}$$

Proof Let $G \in C^1(E, \mathbb{R})$ satisfy (6.1) and set

$$c := \inf_{\sigma \in \Sigma} \sup_{u \in A} G(\sigma(1) u).$$

Then $c \geqslant b$ by (6.2) and $c \leqslant a$ since the identity $\sigma(t) u \equiv u$ is in Σ, so G satisfies $(\text{PS})_c$. We claim that c is a critical value of G. If not, there are $\varepsilon > 0$ and $\eta \in \Sigma$ such that $\eta(1) G^{c+\varepsilon} \subset G^{c-\varepsilon}$ by Lemma 1.3.5. Take a $\sigma \in \Sigma$ such that $\sigma(1) A \subset G^{c+\varepsilon}$ and define $\tilde{\sigma} \in \Sigma$ by

$$\tilde{\sigma}(t) u = \begin{cases} \sigma(2t) u, & 0 \leqslant t \leqslant 1/2 \\ \eta(2t - 1) \sigma(1) u, & 1/2 < t \leqslant 1. \end{cases}$$

Then $\tilde{\sigma}(1) A \subset G^{c-\varepsilon}$, contradicting the definition of c. $\qquad\square$

Remark 6.2.2 Theorem 6.2.1 was proved by Perera and Schechter [122]; see also [120, 121].

6.3 Cohomological index

Next we construct a class of sandwich pairs, based on the cohomological index of Fadell and Rabinowitz [51], applicable to elliptic boundary value problems.

Let us recall the construction and some properties of the cohomological index. Writing the group $\mathbb{Z}_2$ multiplicatively as $\{1, -1\}$, a paracompact $\mathbb{Z}_2$-space is a paracompact space X together with a continuous mapping $\mu : \mathbb{Z}_2 \times X \to X$, called a $\mathbb{Z}_2$-action on X, such that

$$\mu(1, x) = x, \quad -(-x) = x \quad \forall x \in X$$

where $-x := \mu(-1, x)$. The action is fixed-point free if

$$-x \neq x \quad \forall x \in X.$$

A subset A of X is invariant if

$$-A := \{-x : x \in A\} = A,$$

and a map $f : X \to X'$ between two paracompact $\mathbb{Z}_2$-spaces is equivariant if

$$f(-x) = -f(x) \quad \forall x \in X.$$

Two spaces X and X' are equivalent if there is an equivariant homeomorphism $f : X \to X'$. We denote by $\mathcal{F}$ the set of all paracompact free $\mathbb{Z}_2$-spaces, identifying equivalent ones.

A principal $\mathbb{Z}_2$-bundle with paracompact base is a triple $\xi = (E, p, B)$ consisting of an $E \in \mathcal{F}$, called the total space, a paracompact space B, called the base space, and a map $p : E \to B$, called the bundle projection, such that there are

(1) an open covering $\{U_\lambda\}_{\lambda \in \Lambda}$ of B,
(2) for each $\lambda \in \Lambda$, a homeomorphism $\varphi_\lambda : U_\lambda \times \mathbb{Z}_2 \to p^{-1}(U_\lambda)$ satisfying

$$\varphi_\lambda(b, -1) = -\varphi_\lambda(b, 1), \quad p \circ \varphi_\lambda(b, \pm 1) = b \quad \forall b \in B.$$

Then each $p^{-1}(b)$, called a fiber, is some pair $\{e, -e\}$, $e \in E$. A bundle map $f : \xi \to \xi'$ consists of an equivariant map $f : E \to E'$ and a map $\overline{f} : B \to B'$ such that $p' \circ f = \overline{f} \circ p$, i.e. the diagram

$$
\begin{array}{ccc}
E & \xrightarrow{\ f\ } & E' \\
{\scriptstyle p}\downarrow & & \downarrow{\scriptstyle p'} \\
B & \xrightarrow{\ \overline{f}\ } & B'
\end{array}
$$

commutes. Two bundles ξ and ξ' are equivalent if there are bundle maps $f : \xi \to \xi'$ and $f' : \xi' \to \xi$ such that $f' \circ f$ and $f \circ f'$ are the identity bundle maps on ξ and ξ', respectively. We denote by $\mathrm{Prin}_{\mathbb{Z}_2} B$ the set of principal $\mathbb{Z}_2$-bundles over B and $\mathrm{Prin}\,\mathbb{Z}_2$ the set of all principal $\mathbb{Z}_2$-bundles with paracompact base, identifying equivalent ones.

Each $X \in \mathcal{F}$ can be identified with a $\xi \in \mathrm{Prin}\,\mathbb{Z}_2$ as follows. Let $\overline{X} = X/\mathbb{Z}_2$ be the (paracompact) quotient space of $X \in \mathcal{F}$ with each x and $-x$ identified, called the orbit space of X, and $\pi : X \to \overline{X}$ the quotient map. Then

$$\mathcal{P} : \mathcal{F} \to \mathrm{Prin}\,\mathbb{Z}_2, \quad X \mapsto \xi := (X, \pi, \overline{X})$$

is a one-to-one correspondence.

A map $f : B \to B'$ induces a bundle $f^* \xi' = (f^*(E'), p, B) \in \mathrm{Prin}\,\mathbb{Z}_2$, called the pullback, where

$$f^*(E') = \{(b, e') \in B \times E' : f(b) = p'(e')\}, \quad -(b, e') = (b, -e')$$

and

$$p(b, e') = b.$$

Homotopic maps induce equivalent bundles, so for each $\xi' \in \mathrm{Prin}\,\mathbb{Z}_2$, we have the mapping

$$\mathcal{T} : [B, B'] \to \mathrm{Prin}_{\mathbb{Z}_2} B, \quad [f] \mapsto f^* \xi'$$

where $[B, B']$ is the set of homotopy classes of maps from B to B'. For the bundle $\xi' = (S^\infty, \pi, \mathbb{R}P^\infty)$, called the universal principal $\mathbb{Z}_2$-bundle, where S^∞ is the unit sphere in $\mathbb{R}^\infty$, $\mathbb{R}P^\infty$ is the infinite-dimensional real projective space, and π identifies antipodal points $\pm x$, T is a one-to-one correspondence (see Dold [47]). Thus for each $X \in \mathcal{F}$, there is a map $f : \overline{X} \to \mathbb{R}P^\infty$, unique up to homotopy and called the classifying map, such that

$$T([f]) = \mathcal{P}(X).$$

Let $f^* : H^*(\mathbb{R}P^\infty) \to H^*(\overline{X})$ be the induced homomorphism of the Alexander–Spanier cohomology rings. The cohomological index of X is defined by

$$i(X) = \begin{cases} \sup\{k \geqslant 1 : f^*(\omega^{k-1}) \neq 0\}, & X \neq \varnothing \\ 0, & X = \varnothing \end{cases}$$

where $\omega \in H^1(\mathbb{R}P^\infty)$ is the generator of the polynomial ring $H^*(\mathbb{R}P^\infty) = \mathbb{Z}_2[\omega]$.

The index $i : \mathcal{F} \to \mathbb{N} \cup \{0, \infty\}$ has the usual properties of an index theory.

(1) Definiteness: $i(X) = 0$ if and only if $X = \varnothing$.
(2) Monotonicity: If $f : X \to Y$ is an equivariant map, in particular, if $X \subset Y$, then

$$i(X) \leqslant i(Y).$$

Thus, equality holds when f is an equivariant homeomorphism.
(3) Subadditivity: If $X \in \mathcal{F}$ and A, B are closed invariant subsets of X such that $X = A \cup B$, then

$$i(X) \leqslant i(A) + i(B).$$

(4) Continuity: If $X \in \mathcal{F}$ and A is a closed invariant subset of X, then there is a closed invariant neighborhood N of A in X such that

$$i(N) = i(A).$$

(5) Neighborhood of zero: If U is a bounded symmetric neighborhood of 0 in a Banach space E, then

$$i(\partial U) = \dim E.$$

Recall that the suspension SA of a nonempty subset A of a Banach space E is the quotient space of $A \times [-1, 1]$ with $A \times \{1\}$ and $A \times \{-1\}$ collapsed to different points, which can be realized in $E \oplus \mathbb{R}$ as the union of all line segments joining the two points $(0, \pm 1) \in E \oplus \mathbb{R}$ to points of A. The cohomological index also has the following important stability property: If A is a closed symmetric subset of $E \setminus \{0\}$, then

$$i(SA) = i(A) + 1.$$

Let

$$S = \{u \in E : \|u\| = 1\}$$

be the unit sphere in E and

$$\pi : E \setminus \{0\} \to S, \quad u \mapsto \frac{u}{\|u\|}$$

the radial projection onto S. Now let $\mathcal{M}$ be a bounded symmetric subset of $E \setminus \{0\}$ radially homeomorphic to S, i.e. $g = \pi|_{\mathcal{M}} : \mathcal{M} \to S$ is a homeomorphism. Then the radial projection from $E \setminus \{0\}$ onto $\mathcal{M}$ is given by $\pi_{\mathcal{M}} = g^{-1} \circ \pi$. For $A \subset \mathcal{M}$ and $r \geq 0$, we set

$$rA = \{ru : u \in A\}, \qquad \tilde{A} = \pi_{\mathcal{M}}^{-1}(A) \cup \{0\} = \bigcup_{r \geq 0} rA.$$

Theorem 6.3.1 *If A_0, B_0 is a pair of disjoint nonempty closed symmetric subsets of $\mathcal{M}$ such that*

$$i(A_0) = i(\mathcal{M} \setminus B_0) < \infty \tag{6.3}$$

and h is an odd homeomorphism of E such that

$$dist(h(rA_0), h(\tilde{B}_0)) \to \infty \quad as \quad r \to \infty, \tag{6.4}$$

then $A = h(\tilde{A}_0)$, $B = h(\tilde{B}_0)$ form a sandwich pair.

Proof By Theorem 6.2.1, it suffices to verify (6.2), so suppose there is a $\sigma \in \Sigma$ with

$$\sigma(1) A \cap B = \varnothing. \tag{6.5}$$

By (6.4), there is an $R > 1$ such that

$$dist(h(RA_0), B) > \sup_{(u,t) \in E \times [0,1]} \|\sigma(t) u - u\|$$

and hence

$$\sigma(t) h(RA_0) \cap B = \varnothing \quad \forall t \in [0, 1]. \tag{6.6}$$

By (6.5) and (6.6), we can define a map $\eta \in C(A_0 \times [0, 1], E \setminus B)$ by

$$\eta(u, t) = \begin{cases} h((1 - 3t + 3Rt) u), & 0 \leq t \leq 1/3 \\ \sigma(3t - 1) h(Ru), & 1/3 < t \leq 2/3 \\ \sigma(1) h(3(1 - t) Ru), & 2/3 < t \leq 1. \end{cases}$$

Since $\eta|_{A_0 \times \{0\}} = h|_{A_0}$ is odd and $\eta(A_0 \times \{1\})$ is the single point $\sigma(1) h(0)$, η can be extended to an odd map $\tilde{\eta} \in C(SA_0, E \backslash B)$. Then $\pi_{\mathcal{M}} \circ h^{-1} \circ \tilde{\eta}$ is an odd continuous map from SA_0 into $\mathcal{M} \backslash B_0$ and hence

$$i(\mathcal{M} \backslash B_0) \geqslant i(SA_0) = i(A_0) + 1$$

by the monotonicity and the stability of the index, contradicting (6.3). $\qquad \square$

Corollary 6.3.2 *If A_0, B_0 is a pair of disjoint nonempty closed symmetric subsets of S such that*

$$i(A_0) = i(S \backslash B_0) < \infty, \qquad dist(A_0, B_0) > 0,$$

then $A = \pi^{-1}(A_0) \cup \{0\}$, $B = \pi^{-1}(B_0) \cup \{0\}$ form a sandwich pair.

Proof Take h to be the identity in Theorem 6.3.1. Since

$$\text{dist}(r A_0, B) = r \, \text{dist}(A_0, B),$$

it suffices to show that $\text{dist}(A_0, B) > 0$. If not, there are sequences $u_j \in A_0$, $v_j \in B_0$, and $s_j \geqslant 0$ such that $\|u_j - s_j v_j\| \to 0$. Since

$$\|u_j - s_j v_j\| \geqslant \big| \|u_j\| - \|s_j v_j\| \big| = |1 - s_j|,$$

$s_j \to 1$. Then

$$\text{dist}(A_0, B_0) \leqslant \|u_j - v_j\| \leqslant \|u_j - s_j v_j\| + \|(s_j - 1) v_j\|$$

$$= \|u_j - s_j v_j\| + |s_j - 1| \to 0,$$

a contradiction. $\qquad \square$

Corollary 6.3.3 *If $E = N \oplus M$, $u = v + w$ is a direct sum decomposition with $1 \leqslant d := \dim N < \infty$, then N, M form a sandwich pair.*

Proof Take $A_0 = S \cap N$, $B_0 = S \cap M$ in Corollary 6.3.2 and note that $i(A_0) = d$. Since $A_0 \subset S \backslash B_0$ and

$$S \backslash B_0 \to A_0, \qquad u \mapsto \frac{v}{\|v\|}$$

is an odd continuous map, $i(A_0) = i(S \backslash B_0)$ by the monotonicity of the index. $\qquad \square$

Remark 6.3.4 Theorem 6.3.1 and Corollary 6.3.2 were proved by Perera and Schechter in [122] and [121], respectively.

6.4 Semilinear problems

Now we use sandwich pairs based on the eigenspaces of the Laplacian to obtain an existence result for the semilinear elliptic boundary value problem

$$
\begin{cases}
-\Delta u = f(x, u) & \text{in } \Omega \\
u = 0 & \text{on } \partial\Omega
\end{cases}
\tag{6.7}
$$

where Ω is a bounded domain in $\mathbb{R}^n$, $n \geqslant 1$ and f is a Carathéodory function on $\Omega \times \mathbb{R}$ satisfying the subcritical growth condition

$$
|f(x, t)| \leqslant C\left(|t|^{r-1} + 1\right) \quad \forall(x, t) \in \Omega \times \mathbb{R}
\tag{6.8}
$$

for some $r \in (1, 2^*)$ and a constant $C > 0$. Weak solutions of (6.7) coincide with critical points of

$$
G(u) = \int_\Omega |\nabla u|^2 - 2F(x, u), \quad u \in E = H_0^1(\Omega)
$$

where $F(x, t) = \int_0^t f(x, s)\, ds$.

Recall that the Dirichlet spectrum of $-\Delta$ consists of isolated eigenvalues λ_l, $l \geqslant 1$ of finite multiplicities satisfying $0 < \lambda_1 < \lambda_2 < \cdots < \lambda_l < \cdots$. Let E_l be the eigenspace of λ_l,

$$
N_l = \bigoplus_{j=1}^l E_j, \qquad M_l = N_l^\perp.
$$

Then $E = N_l \oplus M_l$, $u = v + w$ is an orthogonal decomposition with respect to both the inner product in E and the $L^2(\Omega)$-inner product, and

$$
\|v\|^2 \leqslant \lambda_l \|v\|_{L^2(\Omega)}^2 \quad \forall v \in N_l,
\tag{6.9}
$$

$$
\|w\|^2 \geqslant \lambda_{l+1} \|w\|_{L^2(\Omega)}^2 \quad \forall w \in M_l.
\tag{6.10}
$$

Let

$$
H(x, t) = 2F(x, t) - tf(x, t).
$$

We have the following theorem.

Theorem 6.4.1 *If*

$$
\lambda_l t^2 - W(x) \leqslant 2F(x, t) \leqslant \lambda_{l+1} t^2 + W(x) \quad \forall(x, t) \in \Omega \times \mathbb{R}
\tag{6.11}
$$

for some l and $W \in L^1(\Omega)$, then (6.7) has a solution in the following cases:

(i) $H(x,t) \leqslant C \left(|t|^\tau + 1 \right)$ *and* $\overline{H}(x) := \limsup\limits_{|t| \to \infty} \dfrac{H(x,t)}{|t|^\tau} < 0,$

(ii) $H(x,t) \geqslant -C \left(|t|^\tau + 1 \right)$ *and* $\underline{H}(x) := \liminf\limits_{|t| \to \infty} \dfrac{H(x,t)}{|t|^\tau} > 0,$

for some $\tau \in [1, 2)$.

Proof By Corollary 6.3.3, N_l, M_l form a sandwich pair, and by (6.11), (6.9), and (6.10),

$$G(v) \leqslant \int_\Omega |\nabla v|^2 - \lambda_l v^2 + W(x) \leqslant \|W\|_{L^1(\Omega)} \quad \forall v \in N_l,$$

$$G(w) \geqslant \int_\Omega |\nabla w|^2 - \lambda_{l+1} w^2 - W(x) \geqslant -\|W\|_{L^1(\Omega)} \quad \forall w \in M_l.$$

It only remains to verify the (PS) condition.

By Lemma 3.1.1, it suffices to show that every (PS) sequence (u_j) is bounded, so suppose $\rho_j := \|u_j\| \to \infty$. Setting $\tilde{u}_j := u_j/\rho_j$ we have $\|\tilde{u}_j\| = 1$, so a renamed subsequence of $(\tilde{u}_j)$ converges to some $\tilde{u}$ weakly in E, strongly in $L^2(\Omega)$, and a.e. in Ω. We have

$$1 - \frac{G(u_j)}{\rho_j^2} = \int_\Omega \frac{2F(x, u_j)}{\rho_j^2} \leqslant \int_\Omega \lambda_{l+1} |\tilde{u}_j|^2 + \frac{W(x)}{\rho_j^2}$$

by (6.11), and passing to the limit gives $\lambda_{l+1} \|\tilde{u}\|_{L^2(\Omega)}^2 \geqslant 1$, so $\tilde{u} \neq 0$. Since $\tau \geqslant 1$,

$$\int_\Omega \frac{H(x, u_j)}{\rho_j^\tau} = \frac{\left(G'(u_j), u_j \right)/2 - G(u_j)}{\rho_j^\tau} \to 0. \tag{6.12}$$

(i) We have $|u_j| = \rho_j |\tilde{u}_j| \to \infty$ and hence

$$\overline{\lim} \frac{H(x, u_j)}{\rho_j^\tau} = \overline{\lim} \frac{H(x, u_j)}{|u_j|^\tau} |\tilde{u}_j|^\tau = \overline{H}(x) |\tilde{u}|^\tau$$

a.e. on $\{\tilde{u} \neq 0\}$, while

$$\frac{H(x, u_j)}{\rho_j^\tau} \leqslant C \left(|\tilde{u}_j|^\tau + \frac{1}{\rho_j^\tau} \right) \to 0$$

a.e. on $\{\tilde{u} = 0\}$, so

$$0 = \overline{\lim} \int_\Omega \frac{H(x, u_j)}{\rho_j^\tau} \leqslant \int_{\tilde{u} \neq 0} \overline{H}(x) |\tilde{u}|^\tau$$

by (6.12) and Fatou's lemma. This is a contradiction since $\overline{H} < 0$ a.e. and $\widetilde{u} \neq 0$.

(*ii*) We have $|u_j| = \rho_j |\widetilde{u}_j| \to \infty$ and hence

$$\underline{\lim} \frac{H(x, u_j)}{\rho_j^\tau} = \underline{\lim} \frac{H(x, u_j)}{|u_j|^\tau} |\widetilde{u}_j|^\tau = \underline{H}(x) |\widetilde{u}|^\tau$$

a.e. on $\{\widetilde{u} \neq 0\}$, while

$$\frac{H(x, u_j)}{\rho_j^\tau} \geq -C \left(|\widetilde{u}_j|^\tau + \frac{1}{\rho_j^\tau} \right) \to 0$$

a.e. on $\{\widetilde{u} = 0\}$, so

$$0 = \underline{\lim} \int_\Omega \frac{H(x, u_j)}{\rho_j^\tau} \geq \int_{\widetilde{u} \neq 0} \underline{H}(x) |\widetilde{u}|^\tau$$

by (6.12) and Fatou's lemma. This is a contradiction since $\underline{H} > 0$ a.e. and $\widetilde{u} \neq 0$. □

6.5 *p*-Laplacian problems

Next we use certain cones as sandwich pairs to extend Theorem 6.4.1 to the *p*-Laplacian problem

$$\begin{cases} -\Delta_p u = f(x, u) & \text{in } \Omega \\ \quad\quad u = 0 & \text{on } \partial\Omega \end{cases} \tag{6.13}$$

where Ω is a bounded domain in $\mathbb{R}^n$, $n \geq 1$, $\Delta_p u = \text{div}\left(|\nabla u|^{p-2} \nabla u \right)$ is the *p*-Laplacian of u, $p \in (1, \infty)$, f is a Carathéodory function on $\Omega \times \mathbb{R}$ satisfying the growth condition (6.8) for some $r \in (1, p^*)$, and

$$p^* = \begin{cases} np/(n-p), & n > p \\ \infty, & n \leq p. \end{cases}$$

Weak solutions of (6.13) coincide with critical points of

$$G(u) = \int_\Omega |\nabla u|^p - p \, F(x, u), \quad u \in E = W_0^{1,\,p}(\Omega)$$

where $F(x, t) = \int_0^t f(x, s) \, ds$ and $W_0^{1,\,p}(\Omega)$ is the usual Sobolev space with the norm

$$\|u\| = \left(\int_\Omega |\nabla u|^p \right)^{1/p}.$$

Eigenvalues of the problem

$$\begin{cases} -\Delta_p u = \lambda |u|^{p-2} u & \text{in } \Omega \\ u = 0 & \text{on } \partial\Omega \end{cases}$$

coincide with critical values of the C^1-functional

$$\Psi(u) = \frac{1}{\displaystyle\int_\Omega |u|^p}, \quad u \in S = \{u \in E : \|u\| = 1\}.$$

Denote by $\mathcal{F}_l$ the class of symmetric subsets M of S with cohomological index $i(M) \geq l$. It was shown in Perera *et al.* [113] (see also Perera [110]) that

$$\lambda_l := \inf_{M \in \mathcal{F}_l} \sup_{u \in M} \Psi(u), \quad l \geq 1$$

is a positive, nondecreasing, and unbounded sequence of eigenvalues, and

$$i(\Psi^{\lambda_l}) = i(S \backslash \Psi_{\lambda_{l+1}}) = l \tag{6.14}$$

when $\lambda_l < \lambda_{l+1}$.

Setting

$$H(x, t) = p F(x, t) - t f(x, t),$$

we prove the following theorem.

Theorem 6.5.1 *If $\lambda_l < \lambda_{l+1}$ and*

$$\lambda_l |t|^p - W(x) \leq p F(x, t) \leq \lambda_{l+1} |t|^p + W(x) \quad \forall (x, t) \in \Omega \times \mathbb{R} \tag{6.15}$$

for some l and $W \in L^1(\Omega)$, then (6.13) has a solution in the following cases:

(i) $H(x, t) \leq C (|t|^\tau + 1)$ and $\overline{H}(x) := \limsup\limits_{|t| \to \infty} \dfrac{H(x, t)}{|t|^\tau} < 0,$

(ii) $H(x, t) \geq -C (|t|^\tau + 1)$ and $\underline{H}(x) := \liminf\limits_{|t| \to \infty} \dfrac{H(x, t)}{|t|^\tau} > 0$

for some $\tau \in [1, p)$.

Proof In view of (6.14) we apply Corollary 6.3.2 with $A_0 = \Psi^{\lambda_l}$, $B_0 = \Psi_{\lambda_{l+1}}$. Since $\|u\|_{L^p(\Omega)} = 1/\Psi(u)^{1/p}$,

$$\|u - v\|_{L^p(\Omega)} \geq \|u\|_{L^p(\Omega)} - \|v\|_{L^p(\Omega)} \geq \frac{1}{\lambda_l^{1/p}} - \frac{1}{\lambda_{l+1}^{1/p}} \quad \forall u \in \Psi^{\lambda_l}, \ v \in \Psi_{\lambda_{l+1}},$$

which together with the Sobolev embedding $E \hookrightarrow L^p(\Omega)$ gives

$$\text{dist}(\Psi^{\lambda_l}, \Psi_{\lambda_{l+1}}) > 0.$$

So $A = \pi^{-1}(\Psi^{\lambda_l}) \cup \{0\}$, $B = \pi^{-1}(\Psi_{\lambda_{l+1}}) \cup \{0\}$ form a sandwich pair.
By (6.15),

$$G(u) \leqslant \int_\Omega |\nabla u|^p - \lambda_l |u|^p + W(x)$$

$$= \|u\|^p \left(1 - \frac{\lambda_l}{\Psi(\pi(u))}\right) + \|W\|_{L^1(\Omega)} \leqslant \|W\|_{L^1(\Omega)} \quad \forall u \in \pi^{-1}(\Psi^{\lambda_l}),$$

$$G(u) \geqslant \int_\Omega |\nabla u|^p - \lambda_{l+1} |u|^p - W(x)$$

$$= \|u\|^p \left(1 - \frac{\lambda_{l+1}}{\Psi(\pi(u))}\right) - \|W\|_{L^1(\Omega)} \geqslant - \|W\|_{L^1(\Omega)} \quad \forall u \in \pi^{-1}(\Psi_{\lambda_{l+1}}).$$

It only remains to verify the (PS) condition, and it suffices to show that every (PS) sequence (u_j) is bounded by a standard argument (see, e.g., Perera *et al.* [113]). If $\rho_j := \|u_j\| \to \infty$, setting $\tilde{u}_j := u_j/\rho_j$ we have $\|\tilde{u}_j\| = 1$, so a renamed subsequence of $(\tilde{u}_j)$ converges to some $\tilde{u}$ weakly in E, strongly in $L^p(\Omega)$, and a.e. in Ω. We have

$$1 - \frac{G(u_j)}{\rho_j^p} = \int_\Omega \frac{p \, F(x, u_j)}{\rho_j^p} \leqslant \int_\Omega \lambda_{l+1} |\tilde{u}_j|^p + \frac{W(x)}{\rho_j^p}$$

by (6.15), and passing to the limit gives $\lambda_{l+1} \|\tilde{u}\|^p_{L^p(\Omega)} \geqslant 1$, so $\tilde{u} \neq 0$. Since $\tau \geqslant 1$,

$$\int_\Omega \frac{H(x, u_j)}{\rho_j^\tau} = \frac{(G'(u_j), u_j)/p - G(u_j)}{\rho_j^\tau} \to 0. \tag{6.16}$$

(*i*) We have $|u_j| = \rho_j |\tilde{u}_j| \to \infty$ and hence

$$\overline{\lim} \frac{H(x, u_j)}{\rho_j^\tau} = \overline{\lim} \frac{H(x, u_j)}{|u_j|^\tau} |\tilde{u}_j|^\tau = \overline{H}(x) |\tilde{u}|^\tau$$

a.e. on $\{\tilde{u} \neq 0\}$, while

$$\frac{H(x, u_j)}{\rho_j^\tau} \leqslant C \left(|\tilde{u}_j|^\tau + \frac{1}{\rho_j^\tau}\right) \to 0$$

a.e. on $\{\tilde{u} = 0\}$, so

$$0 = \overline{\lim} \int_\Omega \frac{H(x, u_j)}{\rho_j^\tau} \leqslant \int_{\tilde{u} \neq 0} \overline{H}(x) |\tilde{u}|^\tau$$

by (6.16) and Fatou's lemma. This is a contradiction since $\overline{H} < 0$ a.e. and $\tilde{u} \neq 0$.

(*ii*) We have $|u_j| = \rho_j |\tilde{u}_j| \to \infty$ and hence

$$\varliminf \frac{H(x, u_j)}{\rho_j^\tau} = \varliminf \frac{H(x, u_j)}{|u_j|^\tau} |\tilde{u}_j|^\tau = \underline{H}(x) |\tilde{u}|^\tau$$

a.e. on $\{\tilde{u} \neq 0\}$, while

$$\frac{H(x, u_j)}{\rho_j^\tau} \geqslant -C \left(|\tilde{u}_j|^\tau + \frac{1}{\rho_j^\tau} \right) \to 0$$

a.e. on $\{\tilde{u} = 0\}$, so

$$0 = \varliminf \int_\Omega \frac{H(x, u_j)}{\rho_j^\tau} \geqslant \int_{\tilde{u} \neq 0} \underline{H}(x) |\tilde{u}|^\tau$$

by (6.16) and Fatou's lemma. This is a contradiction since $\underline{H} > 0$ a.e. and $\tilde{u} \neq 0$. □

Remark 6.5.2 Theorem 6.5.1 was proved by Perera and Schechter [121]. Related results can be found in Arcoya and Orsina [12], Bouchala and Drábek [20], Drábek and Robinson [49], Perera [109], and Perera and Schechter [120].

6.6 Anisotropic systems

Finally we use more general curved sandwich pairs made up of orbits of a certain group action on product spaces to extend Theorem 6.5.1 to the anisotropic system

$$\begin{cases} -\Delta_p u = \nabla_u F(x, u) & \text{in } \Omega \\ u = 0 & \text{on } \partial\Omega \end{cases} \tag{6.17}$$

where Ω is a bounded domain in $\mathbb{R}^n$, $n \geqslant 1$, $\Delta_p u = (\Delta_{p_1} u_1, \ldots, \Delta_{p_m} u_m)$ where $u = (u_1, \ldots, u_m)$ and $p = (p_1, \ldots, p_m)$ with each $p_i \in (1, \infty)$, and $F \in C^1(\Omega \times \mathbb{R}^m)$ with $\nabla_u F = (\partial F/\partial u_1, \ldots, \partial F/\partial u_m)$ satisfying

$$\left| \frac{\partial F}{\partial u_i} \right| \leqslant C \left(\sum_{j=1}^m |u_j|^{r_{ij}-1} + 1 \right) \quad \forall (x, u) \in \Omega \times \mathbb{R}^m \tag{6.18}$$

for some $r_{ij} \in (1, 1 + p_j^*(p_i^* - 1)/p_i^*)$ and a constant $C > 0$.

Let

$$E = E_1 \times \cdots \times E_m = \left\{ u = (u_1, \ldots, u_m) : u_i \in E_i \right\}$$

with the norm

$$\|u\| = \left(\sum_{i=1}^{m} \|u_i\|_i^2 \right)^{1/2},$$

where $\|\cdot\|_i$ denotes the norm in $E_i = W_0^{1,\,p_i}(\Omega)$. Then weak solutions of (6.17) coincide with critical points of

$$G(u) = I(u) - \int_\Omega F(x, u), \quad u \in E$$

where

$$I(u) = \sum_{i=1}^{m} \frac{1}{p_i} \int_\Omega |\nabla u_i|^{p_i}.$$

Let us recall some recent results on eigenvalue problems for systems proved by Perera *et al.* [113]. Define a continuous flow on E, as well as on $\mathbb{R}^m$, by

$$(\alpha, u) \mapsto u_\alpha := (|\alpha|^{1/p_1 - 1} \alpha\, u_1, \ldots, |\alpha|^{1/p_m - 1} \alpha\, u_m), \quad \alpha \in \mathbb{R}.$$

Noting that I satisfies

$$I(u_\alpha) = |\alpha|\, I(u) \quad \forall \alpha \in \mathbb{R},\ u \in E, \tag{6.19}$$

we consider the class of eigenvalue problems

$$\begin{cases} -\Delta_p u = \lambda \nabla_u J(x, u) & \text{in } \Omega \\ u = 0 & \text{on } \partial\Omega \end{cases} \tag{6.20}$$

where $J \in C^1(\Omega \times \mathbb{R}^m)$ is positive somewhere and satisfies

$$J(x, u_\alpha) = |\alpha|\, J(x, u) \quad \forall \alpha \in \mathbb{R},\ (x, u) \in \Omega \times \mathbb{R}^m$$

and the growth condition (6.18) (with J in place of F). Then

$$J(u) = \int_\Omega J(x, u), \quad u \in E$$

also satisfies

$$J(u_\alpha) = |\alpha|\, J(u) \quad \forall \alpha \in \mathbb{R},\ u \in E. \tag{6.21}$$

A typical example is

$$J(x, u) = |u_1|^{r_1} \cdots |u_m|^{r_m} \tag{6.22}$$

where $r_i \in (1, p_i)$ and $\sum_{i=1}^{m} \frac{r_i}{p_i} = 1$.

Let

$$\mathcal{M} = \{u \in E : I(u) = 1\}, \qquad \mathcal{M}^+ = \{u \in \mathcal{M} : J(u) > 0\}.$$

Then $\mathcal{M} \subset E \setminus \{0\}$ is a bounded symmetric C^1-Finsler manifold radially homeomorphic to $S = \{u \in E : \|u\| = 1\}$, $\mathcal{M}^+$ is an open submanifold of $\mathcal{M}$, and positive eigenvalues of (6.20) coincide with critical values of the C^1-functional

$$\Psi(u) = \frac{1}{J(u)}, \quad u \in \mathcal{M}^+.$$

Taking $\alpha = -1$ in (6.21) shows that J, and hence also Ψ, is even. Denote by $\mathcal{F}_l$ the class of symmetric subsets M of $\mathcal{M}^+$ with $i(M) \geqslant l$. It was shown by Perera *et al.* [113] that

$$\lambda_l := \inf_{M \in \mathcal{F}_l} \sup_{u \in M} \Psi(u), \quad l \geqslant 1$$

is a positive, nondecreasing, and unbounded sequence of eigenvalues, and

$$i(\Psi^{\lambda_l}) = i(\mathcal{M}^+ \setminus \Psi_{\lambda_{l+1}}) = l \tag{6.23}$$

when $\lambda_l < \lambda_{l+1}$.

We assume that there are $\tau = (\tau_1, \ldots, \tau_m)$ with $\tau_i \in [1, p_i)$ and $0 < \gamma \leqslant \min_i p_i / \tau_i$ such that setting $u_{\tau,\alpha} = (\alpha^{1/\tau_1} u_1, \ldots, \alpha^{1/\tau_m} u_m)$, $\alpha \geqslant 0$ we have

$$\alpha^\gamma J(x, u) \leqslant J(x, u_{\tau,\alpha}) \quad \forall \alpha \geqslant 1, \ (x, u) \in \Omega \times \mathbb{R}^m. \tag{6.24}$$

In example (6.22) we can take $\gamma = \sum_{i=1}^m \frac{r_i}{\tau_i}$ when this sum is $\leqslant \min_i p_i / \tau_i$. Set

$$H(x, u) = F(x, u) - \sum_{i=1}^m \frac{u_i}{p_i} \frac{\partial F}{\partial u_i}(x, u)$$

and

$$T(u) = \sum_{i=1}^m \frac{1}{\tau_i} |u_i|^{\tau_i}.$$

Theorem 6.6.1 *Under the above hypotheses, if $\lambda_l < \lambda_{l+1}$ and*

$$\lambda_l J(x, u) - W(x) \leqslant F(x, u) \leqslant \lambda_{l+1} J(x, u) + W(x) \quad \forall (x, u) \in \Omega \times \mathbb{R}^m$$
$$\tag{6.25}$$

for some l and $W \in L^1(\Omega)$, then (6.17) has a solution in the following cases:

(i) $H(x, u) \leqslant C(T(u) + 1)$ *and* $\overline{H}(x) := \limsup_{|u| \to \infty} \frac{H(x, u)}{T(u)} < 0$,

(ii) $H(x, u) \geqslant -C(T(u) + 1)$ *and* $\underline{H}(x) := \liminf_{|u| \to \infty} \frac{H(x, u)}{T(u)} > 0$.

Proof In view of (6.23) we apply Theorem 6.3.1 with $A_0 = \Psi^{\lambda_l}$, $B_0 = \Psi_{\lambda_{l+1}} \cup (\mathcal{M} \backslash \mathcal{M}^+)$. Identifying E with $\{\alpha u : u \in \mathcal{M}, \alpha \geqslant 0\}$, define an odd homeomorphism of E by

$$h(\alpha u) = u_\alpha.$$

To see that (6.4) holds, let $\bar{p} = \max_i p_i$. Then for $r \geqslant 1$,

$$\text{dist}(h(r A_0), B) = \inf_{\substack{u \in A_0, \, v \in B_0 \\ s \geqslant 0}} \| u_r - v_s \|$$

$$= \inf_{\substack{u \in A_0, \, v \in B_0 \\ s \geqslant 0}} \left(\sum_{i=1}^m \left\| r^{1/p_i} u_i - s^{1/p_i} v_i \right\|_i^2 \right)^{1/2}$$

$$\geqslant r^{1/\bar{p}} \inf_{\substack{u \in A_0, \, v \in B_0 \\ s \geqslant 0}} \left(\sum_{i=1}^m \left\| u_i - s^{1/p_i} v_i \right\|_i^2 \right)^{1/2}$$

$$= r^{1/\bar{p}} \inf_{\substack{u \in A_0, \, v \in B_0 \\ s \geqslant 0}} \| u - v_s \|$$

$$= r^{1/\bar{p}} \, \text{dist}(A_0, B),$$

so it suffices to show that $\text{dist}(A_0, B) > 0$. If not, there are sequences $u^j \in A_0$, $v^j \in B_0$, and $s_j \geqslant 0$ such that, writing $\tilde{v}^j = v_{s_j}^j$, $\| u^j - \tilde{v}^j \| \to 0$. Then

$$\left| \left\| u_i^j \right\|_i - \left\| \tilde{v}_i^j \right\|_i \right| \leqslant \left\| u_i^j - \tilde{v}_i^j \right\|_i \leqslant \| u^j - \tilde{v}^j \|$$

implies, first that $\left(\left\| \tilde{v}_i^j \right\|_i \right)$ is bounded since $\left(\left\| u_i^j \right\|_i \right)$ is bounded, and then that $\left| \left\| u_i^j \right\|_i^{p_i} - \left\| \tilde{v}_i^j \right\|_i^{p_i} \right| \to 0$ via the elementary inequality

$$|a^p - b^p| \leqslant p \max\{a, b\}^{p-1} |a - b| \quad \forall a, b \geqslant 0, \, p > 1.$$

Since

$$1 - s_j = I(u^j) - s_j I(v^j) = I(u^j) - I(\tilde{v}^j) = \sum_{i=1}^m \frac{1}{p_i} \left(\left\| u_i^j \right\|_i^{p_i} - \left\| \tilde{v}_i^j \right\|_i^{p_i} \right)$$

by (6.19), then $s_j \to 1$. Thus,

$$\text{dist}(A_0, B_0) \leqslant \| u^j - v^j \| \leqslant \| u^j - \tilde{v}^j \| + \| \tilde{v}^j - v^j \|$$

$$= \| u^j - \tilde{v}^j \| + \left(\sum_{i=1}^m \left(s_j^{1/p_i} - 1 \right)^2 \left\| v_i^j \right\|_i^2 \right)^{1/2} \to 0.$$

But $\operatorname{dist}(A_0, B_0) > 0$ since for all $u \in A_0$ and $v \in B_0$,

$$\frac{1}{\lambda_l} - \frac{1}{\lambda_{l+1}} \leqslant J(u) - J(v)$$

$$= \int_\Omega J(x, u) - J(x, v)$$

$$= \int_\Omega \int_0^1 \frac{d}{dt} \left[J(x, t\,u + (1-t)\,v) \right] dt$$

$$= \int_\Omega \int_0^1 \sum_{i=1}^m \frac{\partial J}{\partial u_i}(x, t\,u + (1-t)\,v)\,(u_i - v_i)\,dt$$

$$\leqslant C \int_\Omega \int_0^1 \sum_{i=1}^m \left(\sum_{j=1}^m |t\,u_j + (1-t)\,v_j|^{r_{ij}-1} + 1 \right) |u_i - v_i|\,dt$$

$$\leqslant C \sum_{i=1}^m \left(\sum_{j=1}^m \left(\|u_j\|_{L^{p_j^*}(\Omega)} + \|v_j\|_{L^{p_j^*}(\Omega)} \right)^{r_{ij}-1} \|u_i - v_i\|_{L^{p_i^*}(\Omega)} \right.$$

$$\left. + \|u_i - v_i\|_{L^1(\Omega)} \right) \quad \text{since } (r_{ij} - 1)p_i^*/(p_i^* - 1) < p_j^*$$

$$\leqslant C \sum_{i=1}^m \|u_i - v_i\|_i \qquad \text{since } \mathcal{M} \text{ is bounded}$$

$$\leqslant C \|u - v\|,$$

a contradiction. So

$$A = h(\widetilde{A}_0) = \{ u_\alpha : u \in A_0, \alpha \geqslant 0 \}, \quad B = h(\widetilde{B}_0) = \{ u_\alpha : u \in B_0, \alpha \geqslant 0 \}$$

form a sandwich pair.

Since

$$I(u_\alpha) = \alpha, \quad J(u_\alpha) = \alpha J(u) \quad \forall u \in \mathcal{M}, \alpha \geqslant 0$$

by (6.19) and (6.21), respectively, (6.25) gives

$$G(u_\alpha) \leqslant I(u_\alpha) - \int_\Omega \lambda_l J(x, u_\alpha) - W(x)$$

$$= \alpha (1 - \lambda_l J(u)) + \|W\|_{L^1(\Omega)} \leqslant \|W\|_{L^1(\Omega)} \quad \forall u \in A_0, \alpha \geqslant 0,$$

$$G(u_\alpha) \geqslant I(u_\alpha) - \int_\Omega \lambda_{l+1} J(x, u_\alpha) + W(x)$$

$$= \alpha (1 - \lambda_{l+1} J(u)) - \|W\|_{L^1(\Omega)} \geqslant - \|W\|_{L^1(\Omega)} \quad \forall u \in B_0, \alpha \geqslant 0.$$

It only remains to verify the (PS) condition, and it suffices to show that every (PS) sequence (u^j) is bounded by a standard argument (see, e.g., Perera *et al.* [113]). Noting that

$$T(u_{\tau,\alpha}) = \alpha\, T(u) \quad \forall \alpha \geqslant 0,\ u \in \mathbb{R}^m,$$

if $\rho_j := T(\|u_1^j\|_1, \ldots, \|u_m^j\|_m) \to \infty$, setting $\tilde{u}^j := u_{\tau,1/\rho_j}^j$ we have $T(\|\tilde{u}_1^j\|_1, \ldots, \|\tilde{u}_m^j\|_m) = 1$, so a renamed subsequence of $(\tilde{u}^j)$ converges to some $\tilde{u}$ weakly in E, strongly in $L^{q_1}(\Omega) \times \cdots \times L^{q_m}(\Omega)$ for all $q_i < p_i^*$, and a.e. in $\Omega \times \cdots \times \Omega$. We have

$$\rho_j^\gamma \leqslant C\, I(u^j)$$

for some constant $C > 0$ since $\tau_i\, \gamma \leqslant p_i$ for each i, and

$$\frac{I(u^j) - G(u_j)}{\rho_j^\gamma} = \int_\Omega \frac{F(x, u^j)}{\rho_j^\gamma} \leqslant \int_\Omega \frac{\lambda_{l+1} J(x, u^j) + W(x)}{\rho_j^\gamma}$$

$$\leqslant \lambda_{l+1} \int_\Omega J(x, \tilde{u}^j) + \frac{\|W\|_{L^1(\Omega)}}{\rho_j^\gamma}$$

for all sufficiently large j by (6.25) and (6.24). Combining these inequalities and passing to the limit gives $J(\tilde{u}) > 0$, so $\tilde{u} \neq 0$ since taking $\alpha = 0$ in (6.21) shows that $J(0) = 0$. Since each $\tau_i \geqslant 1$,

$$\int_\Omega \frac{H(x, u^j)}{\rho_j} = \frac{\left(G'(u^j), (u_1^j/p_1, \ldots, u_m^j/p_m) \right) - G(u^j)}{\rho_j} \to 0. \quad (6.26)$$

(i) We have $T(u^j) = \rho_j\, T(\tilde{u}^j) \to \infty$ and hence

$$\overline{\lim}\, \frac{H(x, u^j)}{\rho_j} = \overline{\lim}\, \frac{H(x, u^j)}{T(u^j)}\, T(\tilde{u}^j) = \overline{H}(x)\, T(\tilde{u})$$

a.e. on $\{\tilde{u} \neq 0\}$, while

$$\frac{H(x, u^j)}{\rho_j} \leqslant C \left(T(\tilde{u}^j) + \frac{1}{\rho_j} \right) \to 0$$

a.e. on $\{\tilde{u} = 0\}$, so

$$0 = \overline{\lim} \int_\Omega \frac{H(x, u^j)}{\rho_j} \leqslant \int_{\tilde{u} \neq 0} \overline{H}(x)\, T(\tilde{u})$$

by (6.26) and Fatou's lemma. This is a contradiction since $\overline{H} < 0$ a.e. and $\tilde{u} \neq 0$.

(*ii*) We have $T(u^j) = \rho_j \, T(\tilde{u}^j) \to \infty$ and hence

$$\varliminf \frac{H(x, u^j)}{\rho_j} = \varliminf \frac{H(x, u^j)}{T(u^j)} \, T(\tilde{u}^j) = \underline{H}(x) \, T(\tilde{u})$$

a.e. on $\{\tilde{u} \neq 0\}$, while

$$\frac{H(x, u^j)}{\rho_j} \geq -C \left(T(\tilde{u}^j) + \frac{1}{\rho_j} \right) \to 0$$

a.e. on $\{\tilde{u} = 0\}$, so

$$0 = \varliminf \int_\Omega \frac{H(x, u^j)}{\rho_j} \geq \int_{\tilde{u} \neq 0} \underline{H}(x) \, T(\tilde{u})$$

by (6.26) and Fatou's lemma. This is a contradiction since $\underline{H} > 0$ a.e. and $\tilde{u} \neq 0$.   □

Remark 6.6.2 A result related to Theorem 6.6.1 can be found in Perera and Schechter [122].

Appendix

Sobolev spaces

A.1 Sobolev inequality

Let $\Omega \subset \mathbb{R}^n$ be a bounded open set, and let $C_0^\infty(\Omega)$ denote the set of infinitely differentiable functions on Ω that vanish near the boundary $\partial\Omega$ of Ω. The basic Sobolev inequality is as follows.

Theorem A.1.1 *For each $p \geqslant 1$, $q \geqslant 1$ satisfying*

$$\frac{1}{p} \leqslant \frac{1}{q} + \frac{1}{n},$$

there is a constant $C = C(p, q)$ such that

$$|u|_q \leqslant C \left(|\nabla u|_p + |u|_p \right), \quad u \in C_0^\infty(\Omega),$$

where

$$|u|_q = \left(\int_\Omega |u|^q \, dx \right)^{1/q}, \qquad |\nabla u| = \left(\sum_{k=1}^n \left| \frac{\partial u}{\partial x_k} \right|^2 \right)^{1/2}.$$

If $p > n$, then

$$|u|_\infty \leqslant C \left(|\nabla u|_p + |u|_p \right), \quad u \in C_0^\infty(\Omega),$$

where

$$|u|_\infty = \operatorname*{ess\,sup}_\Omega |u|.$$

A.2 Sobolev spaces

For a nonnegative integer m and $p \geqslant 1$, consider the norm

$$\|u\|_{m,p} = \sum_{|\tau| \leqslant m} |D^\tau u|_p, \quad u \in C_0^\infty(\Omega).$$

An equivalent norm is

$$\left(\sum_{|\tau| \leqslant m} \int_\Omega |D^\tau u|^p \, dx \right)^{1/p}.$$

Theorem A.2.1 *If*

$$p \geqslant 1, \, q \geqslant 1, \, \frac{1}{p} \leqslant \frac{1}{q} + \frac{m}{n},$$

then

$$|u|_q \leqslant C \|u\|_{m,p}, \quad u \in C_0^\infty(\Omega).$$

Proof The theorem is true for $m = 1$ in view of Theorem A.1.1. Assume it is true for $m - 1$. Let q_1 satisfy

$$\frac{1}{q_1} = \frac{1}{p} - \frac{m-1}{n}$$

if $m - 1 < n/p$ and $q_1 > n$ otherwise. In either case,

$$\frac{1}{p} \leqslant \frac{1}{q_1} + \frac{m-1}{n}.$$

By the induction hypothesis,

$$|u|_{q_1} \leqslant C \|u\|_{m-1,p}, \quad u \in C_0^\infty(\Omega).$$

Thus,

$$|D_j u|_{q_1} \leqslant C \|D_j u\|_{m-1,p} \leqslant C \|u\|_{m,p}, \quad u \in C_0^\infty(\Omega).$$

Hence

$$|\nabla u|_{q_1} + |u|_{q_1} \leqslant C \|u\|_{m,p}, \quad u \in C_0^\infty(\Omega).$$

Moreover,

$$\frac{1}{q_1} \leqslant \frac{1}{q} + \frac{1}{n}$$

since

$$\frac{1}{p} - \frac{m-1}{n} \leqslant \frac{1}{q} + \frac{1}{n}.$$

So

$$|u|_q \leqslant C\left(|\nabla u|_{q_1} + |u|_{q_1}\right) \leqslant C\|u\|_{m,p},$$

and the theorem is proved. □

Let $W_0^{m,p}(\Omega)$ be the completion of $C_0^\infty(\Omega)$ with respect to the norm $\|\cdot\|_{m,p}$. What kind of functions are in $W_0^{m,p}(\Omega)$? If $u \in W_0^{m,p}(\Omega)$, then there is a sequence $(u_k) \subset C_0^\infty(\Omega)$ such that

$$\|u_k - u\|_{m,p} \to 0.$$

So

$$\|u_j - u_k\|_{m,p} \to 0.$$

This means that

$$\sum_{|\tau|\leqslant m} |D^\tau u_j - D^\tau u_k|_p \to 0 \quad \text{as} \quad j,k \to \infty.$$

Consequently, for each τ such that $|\tau| \leqslant m$, there is a function $u_\tau \in L^p(\Omega)$ such that

$$|D^\tau u_k - u_\tau|_p \to 0.$$

The function u_τ does not depend on the sequence (u_k), for if $(\hat{u}_k)$ is another sequence converging to u in $W_0^{m,p}(\Omega)$, then

$$\|\hat{u}_k - u_k\|_{m,p} \to 0.$$

This implies that $\hat{u}_\tau = u_\tau$ for each τ. We call u_τ the generalized strong D^τ derivative of u in $L^p(\Omega)$, and denote it by $D^\tau u$. We have the following theorem.

Theorem A.2.2 *Under the hypotheses of Theorem A.2.1,*

$$|u|_q \leqslant C\|u\|_{m,p}, \quad u \in W_0^{m,p}(\Omega). \tag{A.1}$$

Proof For a sequence $(u_k) \subset C_0^\infty(\Omega)$ converging to u in $W_0^{m,p}(\Omega)$, we have by Theorem A.2.1,

$$|u_j - u_k|_q \leqslant C\|u_j - u_k\|_{m,p}.$$

So $u_k \to \hat{u}$ in $L^q(\Omega)$. Since $u_k \to u$ in $L^p(\Omega)$, we must have $\hat{u} = u$. Since

$$|u_k|_q \leqslant C\|u_k\|_{m,p},$$

we have (A.1). □

We also have the following theorem.

Theorem A.2.3 *If $m > n/p$, then*

$$|u|_\infty \leqslant C \|u\|_{m,p}, \quad u \in C_0^\infty(\Omega).$$

Proof If $m - 1 < n/p$, let

$$\frac{1}{q} = \frac{1}{p} - \frac{m-1}{n}.$$

Otherwise, take $q > n$. Then

$$|u|_q \leqslant C \|u\|_{m-1,p}, \quad u \in C_0^\infty(\Omega)$$

in view of Theorem A.2.1. Hence

$$|D_j u|_q \leqslant C \|D_j u\|_{m-1,p} \leqslant C \|u\|_{m,p},$$

implying

$$|\nabla u|_q \leqslant C \|u\|_{m,p}.$$

Since $q > n$, we have

$$|u|_\infty \leqslant C \left(|\nabla u|_q + |u|_q\right) \leqslant C \|u\|_{m,p}.$$

$\square$

Theorem A.2.4 *If $m > n/p$ and $u \in W_0^{m,p}(\Omega)$, then $u \in C(\Omega)$ and*

$$\max_\Omega |u| \leqslant C \|u\|_{m,p}.$$

Proof If (u_k) is a sequence in $C_0^\infty(\Omega)$ converging to u in $W_0^{m,p}(\Omega)$, then

$$|u_j - u_k|_\infty \leqslant C \|u_j - u_k\|_{m,p} \to 0.$$

Hence u_k converges uniformly on Ω to a continuous function $\hat{u}$. Since $u_k \to u$ in $L^p(\Omega)$, we must have $\hat{u} = u$. $\square$

Corollary A.2.5 *If $m - \ell > n/p$, then $W_0^{m,p}(\Omega) \subset C^\ell(\Omega)$ and*

$$\max_{|\tau| \leqslant \ell} \max_\Omega |D^\tau u| \leqslant C \|u\|_{m,p}, \quad u \in W_0^{m,p}(\Omega).$$

Proof We apply Theorem A.2.4 to the derivatives of u up to order ℓ. $\square$

Bibliography

[1] Agarwal, R. P., V. Otero-Espinar, K. Perera, and D. R. Vivero. Basic properties of Sobolev's spaces on time scales. *Adv. Difference Equ.*, pages Art. ID 38121, **14**, 2006.

[2] Ahmad, S., A. C. Lazer, and J. L. Paul. Elementary critical point theory and perturbations of elliptic boundary value problems at resonance. *Indiana Univ. Math. J.*, **25**(10): 933–944, 1976.

[3] Amann, H. and E. Zehnder. Nontrivial solutions for a class of nonresonance problems and applications to nonlinear differential equations. *Ann. Scuola Norm. Sup. Pisa Cl. Sci. (4)*, **7**(4): 539–603, 1980.

[4] Ambrosetti, A. Differential equations with multiple solutions and nonlinear functional analysis. In *Equadiff 82 (Würzburg, 1982)*, volume 1017 of *Lecture Notes in Math.*, pages 10–37. Springer, Berlin, 1983.

[5] Ambrosetti, A. Elliptic equations with jumping nonlinearities. *J. Math. Phys. Sci.*, **18**(1): 1–12, 1984.

[6] Ambrosetti, A., H. Brezis, and G. Cerami. Combined effects of concave and convex nonlinearities in some elliptic problems. *J. Funct. Anal.*, **122**(2): 519–543, 1994.

[7] Ambrosetti, A., J. G. Azorero, and I. Peral. Multiplicity results for some nonlinear elliptic equations. *J. Funct. Anal.*, **137**(1): 219–242, 1996.

[8] Ambrosetti, A. and G. Prodi. *A Primer of Nonlinear Analysis*, volume 34 of *Cambridge Studies in Advanced Mathematics*, pages viii+171. Cambridge University Press, Cambridge, 1995. Corrected reprint of the 1993 original.

[9] Ambrosetti, A. and P. H. Rabinowitz. Dual variational methods in critical point theory and applications. *J. Functional Analysis*, **14**: 349–381, 1973.

[10] Anane, A. Simplicité et isolation de la première valeur propre du p-laplacien avec poids. *C. R. Acad. Sci. Paris Sér. I Math.*, **305**(16): 725–728, 1987.

[11] Anane, A. and Tsouli, N. On the second eigenvalue of the p-Laplacian. In *Nonlinear Partial Differential Equations (Fès, 1994)*, volume 343 of *Pitman Res. Notes Math. Ser.*, pages 1–9. Longman, Harlow, 1996.

[12] Arcoya, D. and L. Orsina. Landesman–Lazer conditions and quasilinear elliptic equations. *Nonlinear Anal.*, **28**(10): 1623–1632, 1997.

[13] Bartolo, P., V. Benci, and D. Fortunato. Abstract critical point theorems and applications to some nonlinear problems with "strong" resonance at infinity. *Nonlinear Anal.*, **7**(9): 981–1012, 1983.

147

[14] Bartsch, T. and S. Li. Critical point theory for asymptotically quadratic functionals and applications to problems with resonance. *Nonlinear Anal.*, **28**(3): 419–441, 1997.

[15] Benci, V. Some applications of the generalized Morse–Conley index. *Confer. Sem. Mat. Univ. Bari*, **218**: 32, 1987.

[16] Benci, V. A new approach to the Morse–Conley theory and some applications. *Ann. Mat. Pura Appl. (4)*, **158**: 231–305, 1991.

[17] Benci, V. Introduction to Morse theory: a new approach. In *Topological Nonlinear Analysis*, volume 15 of *Progr. Nonlinear Differential Equations Appl.*, pages 37–177. Birkhäuser Boston, Boston, MA, 1995.

[18] Benci, V. and P. H. Rabinowitz. Critical point theorems for indefinite functionals. *Invent. Math.*, **52**(3): 241–273, 1979.

[19] Berger, M. S. *Nonlinearity and Functional Analysis*, volume 74 of *Pure and Applied Mathematics*, pages xix+417. Academic Press, 1977.

[20] Bouchala, J. and P. Drábek. Strong resonance for some quasilinear elliptic equations. *J. Math. Anal. Appl.*, **245**(1): 7–19, 2000.

[21] Brezis, H. and L. Nirenberg. Remarks on finding critical points. *Comm. Pure Appl. Math.*, **44**(8–9): 939–963, 1991.

[22] Các, N. P. On nontrivial solutions of a Dirichlet problem whose jumping nonlinearity crosses a multiple eigenvalue. *J. Differential Equations*, **80**(2): 379–404, 1989.

[23] Cambini, A. Sul lemma di M. Morse. *Boll. Un. Mat. Ital. (4)*, **7**: 87–93, 1973.

[24] Castro, A. and A. C. Lazer. Applications of a maximin principle. *Rev. Colombiana Mat.*, **10**: 141–149, 1976.

[25] Cerami, G. An existence criterion for the critical points on unbounded manifolds. *Istit. Lombardo Accad. Sci. Lett. Rend. A*, **112**(2): 332–336, 1979.

[26] Chabrowski, J. *Variational Methods for Potential Operator Equations*, pages x+290. Walter de Gruyter, 1997.

[27] Chang, K. C. and N. Ghoussoub. The Conley index and the critical groups via an extension of Gromoll–Meyer theory. *Topol. Methods Nonlinear Anal.*, **7**(1): 77–93, 1996.

[28] Chang, K. C. Solutions of asymptotically linear operator equations via Morse theory. *Comm. Pure Appl. Math.*, **34**(5): 693–712, 1981.

[29] Chang, K.-C. *Infinite-dimensional Morse Theory and Multiple Solution Problems*, volume 6 of *Progress in Nonlinear Differential Equations and their Applications*, Birkhäuser Boston Inc., Boston, MA, 1993.

[30] Chang, K.-C. *Methods in Nonlinear Analysis*. Springer Monographs in Mathematics, Springer-Verlag, Berlin, 2005.

[31] Cingolani, S. and M. Degiovanni. Nontrivial solutions for p-Laplace equations with right-hand side having p-linear growth at infinity. *Comm. Partial Differential Equations*, **30**(7–9): 1191–1203, 2005.

[32] Corvellec, J.-N. and A. Hantoute. Homotopical stability of isolated critical points of continuous functionals. *Set-Valued Anal.*, **10**(2–3): 143–164, 2002.

[33] Costa, D. G. and E. A. Silva. On a class of resonant problems at higher eigenvalues. *Differential Integral Equations*, **8**(3): 663–671, 1995.

[34] Cuesta, M. On the Fučík spectrum of the Laplacian and the p-Laplacian. In *Proceedings of Seminar in Differential Equations*, Kvilda, Czech Republic, May 29 – June 2, 2000, pages 67–96. Centre of Applied Mathematics, Faculty of Applied Sciences, University of West Bohemia in Pilsen.

[35] Cuesta, M. and J.-P. Gossez. A variational approach to nonresonance with respect to the Fučík spectrum. *Nonlinear Anal.*, **19**(5): 487–500, 1992.

[36] Dacorogna, B. *Direct Methods in the Calculus of Variations*, pages xii+619, Springer, 2008.

[37] Dancer, E. N. On the Dirichlet problem for weakly non-linear elliptic partial differential equations. *Proc. Roy. Soc. Edinburgh Sect. A*, **76**(4): 283–300, 1976/77.

[38] Dancer, E. N. Corrigendum: "On the Dirichlet problem for weakly nonlinear elliptic partial differential equations" [*Proc. Roy. Soc. Edinburgh Sect. A* **76** (1976/77), no. 4, 283–300; MR **58** #17506]. *Proc. Roy. Soc. Edinburgh Sect. A*, **89**(1–2): 15, 1981.

[39] Dancer, E. N. Remarks on jumping nonlinearities. In *Topics in Nonlinear Analysis*, volume 35 of *Progr. Nonlinear Differential Equations Appl.*, pages 101–116. Birkhäuser, Basel, 1999.

[40] Dancer, E. N. Some results for jumping nonlinearities. *Topol. Methods Nonlinear Anal.*, **19**(2): 221–235, 2002.

[41] Dancer, N. and K. Perera. Some remarks on the Fučík spectrum of the p-Laplacian and critical groups. *J. Math. Anal. Appl.*, **254**(1): 164–177, 2001.

[42] de Figueiredo, D. G. and J.-P. Gossez. On the first curve of the Fučík spectrum of an elliptic operator. *Differential Integral Equations*, **7**(5–6): 1285–1302, 1994.

[43] de Paiva, F. O. and E. Massa. Multiple solutions for some elliptic equations with a nonlinearity concave at the origin. *Nonlinear Anal.*, **66**(12): 2940–2946, 2007.

[44] Degiovanni, M. and S. Lancelotti. Linking over cones and nontrivial solutions for p-Laplace equations with p-superlinear nonlinearity. *Ann. Inst. H. Poincaré Anal. Non Linéaire*, **24**(6): 907–919, 2007.

[45] Degiovanni, M. and S. Lancelotti. Linking solutions for p-Laplace equations with nonlinearity at critical growth. *J. Funct. Anal.*, **256**(11): 3643–3659, 2009.

[46] Degiovanni, M., S. Lancelotti, and K. Perera. Nontrivial solutions of p-superlinear p-laplacian problems via a cohomological local splitting. *Commun. Contemp. Math.*, **12**(3): 475–486, 2010.

[47] Dold, A. Partitions of unity in the theory of fibrations. *Ann. of Math. (2)*, **78**: 223–255, 1963.

[48] Drábek, P. *Solvability and Bifurcations of Nonlinear Equations*, volume 264 of *Pitman Research Notes in Mathematics Series*. Longman Scientific & Technical, Harlow, 1992.

[49] Drábek, P. and S. B. Robinson. Resonance problems for the p-Laplacian. *J. Funct. Anal.*, **169**(1): 189–200, 1999.

[50] Esteban, J. R. and J. L. Vázquez. On the equation of turbulent filtration in one-dimensional porous media. *Nonlinear Anal.*, **10**(11): 1303–1325, 1986.

[51] Fadell, E. R. and P. H. Rabinowitz. Generalized cohomological index theories for Lie group actions with an application to bifurcation questions for Hamiltonian systems. *Invent. Math.*, **45**(2): 139–174, 1978.

[52] Fang, F. and S. Liu. Nontrivial solutions of superlinear p-Laplacian equations. *J. Math. Anal. Appl.*, **351**(1): 138–146, 2009.

[53] Fučík, S. Boundary value problems with jumping nonlinearities. *Časopis Pěst. Mat.*, **101**(1): 69–87, 1976.

[54] Gallouët, T. and O. Kavian. Résultats d'existence et de non-existence pour certains problèmes demi-linéaires à l'infini. *Ann. Fac. Sci. Toulouse Math. (5)*, **3**(3–4): 201–246 (1982), 1981.

[55] Ghoussoub, N. Location, multiplicity and Morse indices of min-max critical points. *J. Reine Angew. Math.*, **417**: 27–76, 1991.

[56] Ghoussoub, N. *Duality and Perturbation Methods in Critical Point Theory*, volume 107 of *Cambridge Tracts in Mathematics*. Cambridge University Press, Cambridge, 1993. With appendices by David Robinson.

[57] Gromoll, D. and W. Meyer. On differentiable functions with isolated critical points. *Topology*, **8**: 361–369, 1969.

[58] Guo, Y. and J. Liu. Solutions of p-sublinear p-Laplacian equation via Morse theory. *J. London Math. Soc. (2)*, **72**(3): 632–644, 2005.

[59] Hirano, N. and T. Nishimura. Multiplicity results for semilinear elliptic problems at resonance and with jumping nonlinearities. *J. Math. Anal. Appl.*, **180**(2): 566–586, 1993.

[60] Hofer, H. The topological degree at a critical point of mountain-pass type. *AMS Proceedings of Symposia in Pure Math.*, **45**: 501–509, 1986.

[61] Kelley, J. L. *General Topology*. Springer-Verlag, New York, 1975. Reprint of the 1955 edition [Van Nostrand, Toronto, Ont.], Graduate Texts in Mathematics, No. 27.

[62] Krasnosel'skii, M. A. *Topological Methods in the Theory of Nonlinear Integral Equations*. Translated by A. H. Armstrong; translation edited by J. Burlak. A Pergamon Press Book. The Macmillan Co., New York, 1964.

[63] Kryszewski, W. and A. Szulkin. An infinite-dimensional Morse theory with applications. *Trans. Amer. Math. Soc.*, **349**(8): 3181–3234, 1997.

[64] Kuiper, N. H. C^1-equivalence of functions near isolated critical points. *Ann. of Math.*, **69**: 199–218, 1972.

[65] Lancelotti, S. Existence of nontrivial solutions for semilinear problems with strictly differentiable nonlinearity. *Abstr. Appl. Anal.*, pages Art. ID 62458, 14, 2006.

[66] Lazer, A. C. and P. J. McKenna. Critical point theory and boundary value problems with nonlinearities crossing multiple eigenvalues. II. *Comm. Partial Differential Equations*, **11**(15): 1653–1676, 1986.

[67] Lazer, A. C. and S. Solimini. Nontrivial solutions of operator equations and Morse indices of critical points of min-max type. *Nonlinear Anal.*, **12**(8): 761–775, 1988.

[68] Lazer, A. Introduction to multiplicity theory for boundary value problems with asymmetric nonlinearities. In *Partial Differential Equations (Rio de Janeiro, 1986)*, volume 1324 of *Lecture Notes in Math.*, pages 137–165. Springer, Berlin, 1988.

[69] Li, C., S. Li, and Z. Liu. Existence of type (II) regions and convexity and concavity of potential functionals corresponding to jumping nonlinear problems. *Calc. Var. Partial Differential Equations*, **32**(2): 237–251, 2008.

[70] Li, S. J. and J. Q. Liu. Morse theory and asymptotic linear Hamiltonian system. *J. Differential Equations*, **78**(1): 53–73, 1989.

[71] Li, S. J. and J. Q. Liu. Nontrivial critical points for asymptotically quadratic function. *J. Math. Anal. Appl.*, **165**(2): 333–345, 1992.

[72] Li, S. J. and M. Willem. Applications of local linking to critical point theory. *J. Math. Anal. Appl.*, **189**(1): 6–32, 1995.

[73] Li, S. and J. Q. Liu. Computations of critical groups at degenerate critical point and applications to nonlinear differential equations with resonance. *Houston J. Math.*, **25**(3): 563–582, 1999.

[74] Li, S., K. Perera, and J. Su. Computation of critical groups in elliptic boundary-value problems where the asymptotic limits may not exist. *Proc. Roy. Soc. Edinburgh Sect. A*, **131**(3): 721–732, 2001.

[75] Li, S. and Z.-Q. Wang. Mountain pass theorem in order intervals and multiple solutions for semilinear elliptic Dirichlet problems. *J. Anal. Math.*, **81**: 373–396, 2000.

[76] Li, S. and M. Willem. Multiple solutions for asymptotically linear boundary value problems in which the nonlinearity crosses at least one eigenvalue. *NoDEA Nonlinear Differential Equations Appl.*, **5**(4): 479–490, 1998.

[77] Li, S., S. Wu, and H.-S. Zhou. Solutions to semilinear elliptic problems with combined nonlinearities. *J. Differential Equations*, **185**(1): 200–224, 2002.

[78] Li, S. and Z. Zhang. Multiple solutions theorems for semilinear elliptic boundary value problems with resonance at infinity. *Discrete Contin. Dynam. Systems*, **5**(3): 489–493, 1999.

[79] Li, S. and W. Zou. The computations of the critical groups with an application to elliptic resonant problems at a higher eigenvalue. *J. Math. Anal. Appl.*, **235**(1): 237–259, 1999.

[80] Lindqvist, P. On the equation $\operatorname{div}\left(|\nabla u|^{p-2}\nabla u\right) + \lambda|u|^{p-2}u = 0$. *Proc. Amer. Math. Soc.*, **109**(1): 157–164, 1990.

[81] Lindqvist, P. Addendum: "On the equation $\operatorname{div}(|\nabla u|^{p-2}\nabla u) + \lambda|u|^{p-2}u = 0$" [*Proc. Amer. Math. Soc.* 109 (1990), no. 1, 157–164; MR 90h:35088]. *Proc. Amer. Math. Soc.*, **116**(2): 583–584, 1992.

[82] Liu, J. Q. The Morse index of a saddle point. *Systems Sci. Math. Sci.*, **2**(1): 32–39, 1989.

[83] Liu, J. Q. and S. J. Li. An existence theorem for multiple critical points and its application. *Kexue Tongbao (Chinese)*, **29**(17): 1025–1027, 1984.

[84] Liu, S. and S. Li. Existence of solutions for asymptotically 'linear' p-Laplacian equations. *Bull. London Math. Soc.*, **36**(1): 81–87, 2004.

[85] Ljusternik, L. and L. Schnirelmann. *Methodes Topologique dans les Problémes Variationnels*. Hermann and Cie, Paris, 1934.

[86] Magalhães, C. A. Semilinear elliptic problem with crossing of multiple eigenvalues. *Comm. Partial Differential Equations*, **15**(9): 1265–1292, 1990.

[87] Margulies, C. A. and W. Margulies. An example of the Fučik spectrum. *Nonlinear Anal.*, **29**(12): 1373–1378, 1997.

[88] Marino, A. and G. Prodi. Metodi perturbativi nella teoria di Morse. *Boll. Un. Mat. Ital. (4)*, **11**(3, suppl.): 1–32, 1975. Collection of articles dedicated to Giovanni Sansone on the occasion of his eighty-fifth birthday.

[89] Marino, A. and G. Prodi. La teoria di Morse per gli spazi di Hilbert. *Rend. Sem. Mat. Univ. Padova*, **41**: 43–68, 1968.

[90] Mawhin, J. and M. Willem. On the generalized Morse lemma. *Bull. Soc. Math. Belg. Sér. B*, **37**(2): 23–29, 1985.

[91] Mawhin, J. and M. Willem. *Critical Point Theory and Hamiltonian Systems*, volume 74 of *Applied Mathematical Sciences*. Springer-Verlag, New York, 1989.

[92] Medeiros, E. and K. Perera. Multiplicity of solutions for a quasilinear elliptic problem via the cohomological index. *Nonlinear Anal.*, **71**(9): 3654–3660, 2009.

[93] Milnor, J. *Morse Theory*, pages vi+153. Princeton University Press, 1963.

[94] Moroz, V. On the Morse critical groups for indefinite sublinear elliptic problems. *Nonlinear Anal.*, **52**(5): 1441–1453, 2003.

[95] Morse, M. Relations between the critical points of a real function of n independent variables. *Trans. Amer. Math. Soc.*, **27**(3): 345–396, 1925.

[96] Motreanu, D. and K. Perera. Multiple nontrivial solutions of Neumann p-Laplacian systems. *Topol. Methods Nonlinear Anal.*, **34**(1): 41–48, 2009.

[97] Ni, W. M. Some minimax principles and their applications in nonlinear elliptic equations. *J. Analyse Math.*, **37**: 248–275, 1980.

[98] Nirenberg, L. *Topics in Nonlinear Functional Analysis*, pages viii+259, American Mathematical Society, 1974. With a chapter by E. Zehnder, Notes by R. A. Artino.

[99] Nirenberg, L. Variational and topological methods in nonlinear problems. *Bull. Amer. Math. Soc. (N.S.)*, **4**(3): 267–302, 1981.

[100] Padial, J. F., P. Takáč, and L. Tello. An antimaximum principle for a degenerate parabolic problem. In *Ninth International Conference Zaragoza-Pau on Applied Mathematics and Statistics*, volume 33 of *Monogr. Semin. Mat. García Galdeano*, pages 433–440. Prensas Univ. Zaragoza, Zaragoza, 2006.

[101] Palais, R. S. and S. Smale. A generalized Morse theory. *Bull. Amer. Math. Soc.*, **70**: 165–172, 1964.

[102] Palais, R. S. Morse theory on Hilbert manifolds. *Topology*, **2**: 299–340, 1963.

[103] Palais, R. S. Critical point theory and the minimax principle. In *Global Analysis (Proc. Sympos. Pure Math., Vol. XV, Berkeley, Calif, 1968)*, pages 185–212. Amer. Math. Soc., Providence, RI, 1970.

[104] Perera, K. Critical groups of pairs of critical points produced by linking subsets. *J. Differential Equations*, **140**(1): 142–160, 1997.

[105] Perera, K. Multiplicity results for some elliptic problems with concave nonlinearities. *J. Differential Equations*, **140**(1): 133–141, 1997.

[106] Perera, K. Critical groups of critical points produced by local linking with applications. *Abstr. Appl. Anal.*, **3**(3–4): 437–446, 1998.

[107] Perera, K. Homological local linking. *Abstr. Appl. Anal.*, **3**(1–2):181–189, 1998.

[108] Perera, K. Applications of local linking to asymptotically linear elliptic problems at resonance. *NoDEA Nonlinear Differential Equations Appl.*, **6**(1): 55–62, 1999.

[109] Perera, K. One-sided resonance for quasilinear problems with asymmetric nonlinearities. *Abstr. Appl. Anal.*, **7**(1): 53–60, 2002.

[110] Perera, K. Nontrivial critical groups in p-Laplacian problems via the Yang index. *Topol. Methods Nonlinear Anal.*, **21**(2): 301–309, 2003.

[111] Perera, K. Nontrivial solutions of p-superlinear p-Laplacian problems. *Appl. Anal.*, **82**(9): 883–888, 2003.

[112] Perera, K. *p*-superlinear problems with jumping nonlinearities. In *Nonlinear Analysis and Applications: to V. Lakshmikantham on his 80th Birthday. Vol. 1, 2*, pages 823–829. Kluwer Acad. Publ., Dordrecht, 2003.

[113] Perera, K., R. P. Agarwal, and D. O'Regan. *Morse Theoretic Aspects of p-Laplacian Type Operators*, volume 161 of *Mathematical Surveys and Monographs*. American Mathematical Society, Providence, RI, 2010.

[114] Perera, K. and M. Schechter. Morse index estimates in saddle point theorems without a finite-dimensional closed loop. *Indiana Univ. Math. J.*, **47**(3): 1083–1095, 1998.

[115] Perera, K. and M. Schechter. Type II regions between curves of the Fučik spectrum and critical groups. *Topol. Methods Nonlinear Anal.*, **12**(2): 227–243, 1998.

[116] Perera, K. and M. Schechter. A generalization of the Amann-Zehnder theorem to nonresonance problems with jumping nonlinearities. *NoDEA Nonlinear Differential Equations Appl.*, **7**(4): 361–367, 2000.

[117] Perera, K. and M. Schechter. The Fučík spectrum and critical groups. *Proc. Amer. Math. Soc.*, **129**(8): 2301–2308 (electronic), 2001.

[118] Perera, K. and M. Schechter. Critical groups in saddle point theorems without a finite dimensional closed loop. *Math. Nachr.*, **243**: 156–164, 2002.

[119] Perera, K. and M. Schechter. Double resonance problems with respect to the Fučík spectrum. *Indiana Univ. Math. J.*, **52**(1): 1–17, 2003.

[120] Perera, K. and M. Schechter. Sandwich pairs in *p*-Laplacian problems. *Topol. Methods Nonlinear Anal.*, **29**(1): 29–34, 2007.

[121] Perera, K. and M. Schechter. Flows and critical points. *NoDEA Nonlinear Differential Equations Appl.*, **15**(4–5): 495–509, 2008.

[122] Perera, K. and M. Schechter. Sandwich pairs for *p*-Laplacian systems. *J. Math. Anal. Appl.*, **358**(2): 485–490, 2009.

[123] Perera, K. and A. Szulkin. *p*-Laplacian problems where the nonlinearity crosses an eigenvalue. *Discrete Contin. Dyn. Syst.*, **13**(3): 743–753, 2005.

[124] Pitcher, E. Inequalities of critical point theory. *Bull. Amer. Math. Soc.*, **64**(1): 1–30, 1958.

[125] Qi, G. J. Extension of Mountain Pass Lemma. *Kexue Tongbao (English Ed.)*, **32**(12): 798–801, 1987.

[126] Rabinowitz, P. H. Variational methods for nonlinear eigenvalue problems. In *Eigenvalues of Non-Linear Problems*, pages 139–195. Springer-Verlag, Berlin, 1974.

[127] Rabinowitz, P. H. Some critical point theorems and applications to semilinear elliptic partial differential equations. *Ann. Scuola Norm. Sup. Pisa Cl. Sci. (4)*, **5**(1): 215–223, 1978.

[128] Rabinowitz, P. H. Some minimax theorems and applications to nonlinear partial differential equations. In *Nonlinear Analysis (Collection of Papers in Honor of Erich H. Rothe)*, pages 161–177. Academic Press, New York, 1978.

[129] Rabinowitz, P. H. *Minimax Methods in Critical Point Theory with Applications to Differential Equations*, volume 65 of *CBMS Regional Conference Series in Mathematics*. Published for the Conference Board of the Mathematical Sciences, Washington, DC, 1986.

[130] Ramos, M. and L. Sanchez. Homotopical linking and Morse index estimates in min-max theorems. *Manuscripta Math.*, **87**(3): 269–284, 1995.

[131] Ribarska, N. K., Ts. Y. Tsachev, and M. I. Krastanov. A saddle point theorem without a finite-dimensional closed loop. *C. R. Acad. Bulgare Sci.*, **51**(11–12): 13–16, 1998.

[132] Rothe, E. H. Some remarks on critical point theory in Hilbert space. In *Nonlinear Problems (Proc. Sympos., Madison, Wis., (1962)*, pages 233–256. University of Wisconsin Press, Madison, WI, 1963.

[133] Rothe, E. H. Some remarks on critical point theory in Hilbert space (continuation). *J. Math. Anal. Appl.*, **20**: 515–520, 1967.

[134] Rothe, E. H. On continuity and approximation questions concerning critical Morse groups in Hilbert space. In *Symposium on Infinite-Dimensional Topology (Louisiana State Univ., Baton Rouge, La., 1967)*, pages 275–295. *Ann. of Math. Studies*, No. 69. Princeton University Press, Princeton, NJ, 1972.

[135] Rothe, E. H. Morse theory in Hilbert space. *Rocky Mountain J. Math.*, **3**: 251–274, 1973. Rocky Mountain Consortium Symposium on Nonlinear Eigenvalue Problems (Santa Fe, NM, 1971).

[136] Ruf, B. On nonlinear elliptic problems with jumping nonlinearities. *Ann. Mat. Pura Appl. (4)*, **128**: 133–151, 1981.

[137] Schechter, M. A generalization of the saddle point method with applications. *Ann. Polon. Math.*, **57**(3): 269–281, 1992.

[138] Schechter, M. New saddle point theorems. In *Generalized Functions and Their Applications (Varanasi, 1991)*, pages 213–219. Plenum, New York, 1993.

[139] Schechter, M. Splitting subspaces and saddle points. *Appl. Anal.*, **49**(1–2): 33–48, 1993.

[140] Schechter, M. The Fučík spectrum. *Indiana Univ. Math. J.*, **43**(4): 1139–1157, 1994.

[141] Schechter, M. Bounded resonance problems for semilinear elliptic equations. *Nonlinear Anal.*, **24**(10): 1471–1482, 1995.

[142] Schechter, M. New linking theorems. *Rend. Sem. Mat. Univ. Padova*, **99**: 255–269, 1998.

[143] Schechter, M. *Linking Methods in Critical Point Theory*. Birkhäuser Boston Inc., Boston, MA, 1999.

[144] Schechter, M. *An Introduction to Nonlinear Analysis*, volume 95 of *Cambridge Studies in Advanced Mathematics*. Cambridge University Press, Cambridge, 2004.

[145] Schechter, M. Sandwich pairs. In *Proc. Conf. Differential & Difference Equations and Applications*, pages 999–1007. Hindawi Publ. Corp., New York, 2006.

[146] Schechter, M. Sandwich pairs in critical point theory. *Trans. Amer. Math. Soc.*, **360**(6): 2811–2823, 2008.

[147] Schechter, M. *Minimax Systems and Critical Point Theory*. Birkhäuser Boston Inc., Boston, MA, 2009.

[148] Schechter, M. and K. Tintarev. Pairs of critical points produced by linking subsets with applications to semilinear elliptic problems. *Bull. Soc. Math. Belg. Sér. B*, **44**(3): 249–261, 1992.

[149] Schwartz, J. T. *Nonlinear Functional Analysis*, pages vii+236. Gordon and Breach, 1969. Notes by H. Fattorini, R. Nirenberg and H. Porta, with an additional chapter by Hermann Karcher.

[150] Silva, E. A. de B. e. Linking theorems and applications to semilinear elliptic problems at resonance. *Nonlinear Anal.*, **16**(5): 455–477, 1991.

[151] Smale, S. Morse theory and a non-linear generalization of the Dirichlet problem. *Ann. of Math. (2)*, **80**: 382–396, 1964.

[152] Solimini, S. Morse index estimates in min-max theorems. *Manuscripta Math.*, **63**(4): 421–453, 1989.

[153] Struwe, M. *Variational Methods*, volume 34 of *Ergebnisse der Mathematik und ihrer Grenzgebiete. 3. Folge. A Series of Modern Surveys in Mathematics [Results in Mathematics and Related Areas. 3rd Series. A Series of Modern Surveys in Mathematics]*. Springer-Verlag, Berlin, fourth edition, 2008.

[154] Su, J. and C. Tang. Multiplicity results for semilinear elliptic equations with resonance at higher eigenvalues. *Nonlinear Anal.*, **44**(3, Ser. A: Theory Methods): 311–321, 2001.

[155] Szulkin, A. Cohomology and Morse theory for strongly indefinite functionals. *Math. Z.*, **209**(3): 375–418, 1992.

[156] Tanaka, M. On the existence of a non-trivial solution for the p-Laplacian equation with a jumping nonlinearity. *Tokyo J. Math.*, **31**(2): 333–341, 2008.

[157] Wang, Z. Q. A note on the second variation theorem. *Acta Math. Sinica*, **30**(1): 106–110, 1987.

[158] Willem, M. *Minimax Theorems*, volume 24 in *Progress in Nonlinear Differential Equations and their Applications*. Birkhäuser Boston Inc., Boston, MA, 1996.

[159] Wu, S.-p. and H. Yang. A class of resonant elliptic problems with sublinear nonlinearity at origin and at infinity. *Nonlinear Anal.*, **45**(7, Ser. A: Theory Methods): 925–935, 2001.

[160] Yang, C.-T. On theorems of Borsuk-Ulam, Kakutani-Yamabe-Yujobô and Dyson. I. *Ann. of Math. (2)*, **60**: 262–282, 1954.

[161] Zeidler, E. *Nonlinear Functional Analysis and its Applications. III*, pages xxii+662, World Publishing Corporation, 1985. Translated from the German by Leo F. Boron.

[162] Zou, W. and J. Q. Liu. Multiple solutions for resonant elliptic equations via local linking theory and Morse theory. *J. Differential Equations*, **170**(1): 68–95, 2001.

[163] Zou, W. and M. Schechter. *Critical Point Theory and its Applications*. Springer, New York, 2006.

Index

Printed in the United States
by Baker & Taylor Publisher Services